1996

BOOKS BY JOHN BROCKMAN

AS AUTHOR:

By the Late John Brockman
37
Afterwords

AS EDITOR:

About Bateson
Speculations
Doing Science
Ways of Knowing
Creativity
How Things Are

The

THIRD
CULTURE

by John Brockman

A TOUCHSTONE BOOK

Published by Simon & Schuster

TOUCHSTONE
Rockefeller Center
1230 Avenue of the Americas
New York, NY 10020

First Touchstone Edition 1996

TOUCHSTONE and colophon are registered trademarks
of Simon & Schuster Inc.

Designed by Liney Li

Manufactured in the United States of America

10   9   8   7   6   5   4   3   2   1

10   9   8   7   6   5   4   3   2   Pbk.

Library of Congress Cataloging-in-Publication Data
Brockman, John, date.
    The third culture : beyond the scientific revolution / by John
Brockman.
        p.   cm.
    Includes bibliographical references (p.).
    1. Scientists—20th century—Interviews.   2. Intellectuals—
20th century—Interviews.   3. Science and the humanities.   I. Title.
Q141.B76   1995
500—dc20   95-83
                                                              CIP

ISBN: 0-684-80359-3
ISBN: 0-684-82344-6 (Pbk.)

CONTENTS

Chapter 2.
STEPHEN JAY GOULD
"The Pattern of Life's History"

There is no progress in evolution. The fact of evolutionary change through time doesn't represent progress as we know it. Progress isn't inevitable. Much of evolution is downward in terms of morphological complexity, rather than upward. We're not marching toward some greater thing.

51

Chapter 3.
RICHARD DAWKINS
"A Survival Machine"

It rapidly became clear to me that the most imaginative way of looking at evolution, and the most inspiring way of teaching it, was to say that it's all about the genes. It's the genes that, for their own good, are manipulating the bodies they ride about in. The individual organism is a survival machine for its genes.

74

Chapter 4.
BRIAN GOODWIN
"Biology Is Just a Dance"

The "new" biology is biology in the form of an exact science of complex systems concerned with dynamics and emergent order. Then everything in biology changes. Instead of the metaphors of conflict, competition, selfish genes, climbing peaks in fitness landscapes, what you get is evolution as a dance. It has no goal. As Stephen Jay Gould says, it has no purpose, no progress, no sense of direction. It's a dance through morphospace, the space of the forms of organisms.

96

Chapter 5.

STEVE JONES

"Why Is There So Much Genetic Diversity?"

We have the beginnings of an answer as to why, in some places, one snail species is so variable, but we have no real idea why in any species anywhere at any time no two individuals are identical. That's an essential question of evolution. All others flow from that.

111

Chapter 6.

NILES ELDREDGE

"A Battle of Words"

Species are real entities, spatiotemporally bounded, and they're information entities. Other kinds of entities do things. Ecological populations, for example, have niches; they function. Species don't function that way. They don't do things; they are, instead, information repositories. A species is not like an organism at all, but it's nonetheless a kind of entity that plays an important role in the evolutionary process.

119

Chapter 7.

LYNN MARGULIS

"Gaia Is a Tough Bitch"

How did the eukaryotic cell appear? Probably it was an invasion of predators, at the outset. It may have started when one sort of squirming bacterium invaded another—seeking food, of course. But certain invasions evolved into truces; associations once ferocious became benign. When swimming bacterial would-be invaders took up residence inside their sluggish hosts, this joining of forces created a new whole that was, in effect, far greater than the sum of its parts: faster swimmers capable of moving huge quantities of genes evolved. Some of these newcomers were uniquely competent in the evolutionary struggle. Further bacterial associations were added on, as the modern cell evolved.

129

Chapter 8.

MARVIN MINSKY
"Smart Machines"

The brain is . . . a great jury-rigged combination of many gadgets to do different things, with additional gadgets to correct their deficiencies, and yet more accessories to intercept their various bugs and undesirable interactions—in short, a great mess of assorted mechanisms that barely manage to get the job done.

152

Chapter 9.

ROGER SCHANK
"Information Is Surprises"

Information is surprises. We all expect the world to work out in certain ways, but when it does, we're bored. What makes something worth knowing is organized around the concept of expectation failure. Scripts are interesting not when they work but when they fail.

167

Chapter 10.

DANIEL C. DENNETT
"Intuition Pumps"

The idea of consciousness as a virtual machine is a nice intuition pump. It takes a while to set up, because a lot of the jargon of artificial intelligence and computer science is unfamiliar to philosophers or other people. But if you have the patience to set some of these ideas up, then you can say, "Hey! Try thinking about the idea that what we have in our heads is software. It's a virtual machine, in the same way that a word processor is a virtual machine." Suddenly, bells start ringing, and people start seeing things from a slightly different perspective.

181

Chapter 11.

NICHOLAS HUMPHREY
"The Thick Moment"

What is it like to be ourselves? How can a piece of matter which is a human be-ing be the basis for the experience each one of us recognizes as what it's like to be us? How can a human body and a human brain also be a human mind?

198

Chapter 12.

FRANCISCO VARELA
"The Emergent Self"

Why do emergent selves, virtual identities, pop up all over the place, creating worlds, whether at the mind/body level, the cellular level, or the transorganism level? This phenomenon is something so productive that it doesn't cease creat-ing entirely new realms: life, mind, and societies. Yet these emergent selves are based on processes so shifty, so ungrounded, that we have an apparent paradox between the solidity of what appears to show up and its groundlessness. That, to me, is the key and eternal question.

209

Chapter 13.

STEVEN PINKER
"Language Is a Human Instinct"

I call language an "instinct," an admittedly quaint term for what other cognitive scientists have called a mental organ, a faculty, or a module. Language is a com-plex, specialized skill, which develops in the child spontaneously without con-scious effort or formal instruction, is deployed without awareness of its underlying logic, is qualitatively the same in every individual, and is distinct from more general abilities to process information or behave intelligently.

223

Chapter 14.

ROGER PENROSE
"Consciousness Involves Noncomputable Ingredients"

My present view is that the brain isn't exactly a quantum computer. Quantum actions are important in the way the brain works, but the noncomputational actions occur at the bridge from the quantum to the classical level, and that bridge is beyond our present understanding of quantum mechanics.

239

PART THREE: QUESTIONS OF ORIGINS

Chapter 15.

MARTIN REES
"An Ensemble of Universes"

Cosmology is exciting to the public because it's clearly fundamental, and this is a rather special time in the subject. For the first time, it's become a part of mainstream science, and we can address questions about the origin of the universe.

262

Chapter 16.

ALAN GUTH
"A Universe in Your Backyard"

One of the most amazing features of the inflationary-universe model is that it allows the universe to evolve from something that's initially incredibly small. Something on the order of twenty pounds of matter is all it seems to take to start off a universe. . . . It becomes very tempting to ask whether, in principle, it's possible to create a universe in the laboratory—or a universe in your backyard—by man-made processes.

276

Chapter 17.
LEE SMOLIN
"A Theory of the Whole Universe"

What is space and what is time? This is what the problem of quantum gravity is about. In general relativity, Einstein gave us not only a theory of gravity but a theory of what space and time are—a theory that overthrew the previous Newtonian conception of space and time. The problem of quantum gravity is how to combine the understanding of space and time we have from relativity theory with the quantum theory, which also tells us something essential and deep about nature.

286

Chapter 18.
PAUL DAVIES
"The Synthetic Path"

My personal belief is that biologists tend to be uncompromising and reductionistic because they're still feeling somewhat insecure with their basic dogma, whereas physicists have three hundred years of secure foundation for their subject, so they can afford to be a bit more freewheeling in their speculation about these complex systems.

303

PART FOUR: WHAT WAS DARWIN'S ALGORITHM?

Chapter 19.
MURRAY GELL-MANN
"Plectics"

To refer to the subject on which some of us now work as "complexity" seems to me to distort the nature of what we do, because the simplicity of the underlying rules is a critical feature of the whole enterprise. Therefore what I like to say is that the subject consists of the study of simplicity, complexity of various kinds, and complex adaptive systems, with some consideration of complex nonadaptive systems as well.

316

Chapter 20.

STUART KAUFFMAN
"Order for Free"

What kinds of complex systems can evolve by accumulation of successive useful variations? Does selection by itself achieve complex systems able to adapt? Are there lawful properties characterizing such complex systems? The overall answer may be that complex systems constructed so that they're on the boundary between order and chaos are those best able to adapt by mutation and selection.

333

Chapter 21.

CHRISTOPHER G. LANGTON
"A Dynamical Pattern"

Physics has largely been the science of necessity, uncovering the fundamental laws of nature and what must be true given those laws. Biology, on the other hand, is the science of the possible, investigating processes that are possible, given those fundamental laws, but not necessary. Biology is consequently much harder than physics but also infinitely richer in its potential, not just for understanding life and its history but for understanding the universe and its future. The past belongs to physics, but the future belongs to biology.

344

Chapter 22.

J. DOYNE FARMER
"The Second Law of Organization"

Many of us believe that self-organization is a general property—certainly of the universe, and even more generally of mathematical systems that might be called "complex adaptive systems." Complex adaptive systems have the property that if you run them—by just letting the mathematical variable of "time" go forward—they'll naturally progress from chaotic, disorganized, undifferentiated, independent states to organized, highly differentiated, and highly interdependent states.

359

Chapter 23.
W. DANIEL HILLIS
"Close to the Singularity"

We're analogous to the single-celled organisms when they were turning into multicellular organisms. We're the amoebas, and we can't quite figure out what the hell this thing is that we're creating. We're right at that point of transition, and there's something coming along after us.

ACKNOWLEDGMENTS

I first published a brief essay about the idea of an emerging third culture in September 1991, in my newsletter *EDGE* (#3). An extended version of the essay was later published by the *Los Angeles Times*, *The New Statesman*, and the Copenhagen daily newspaper *Information*.

Several people made helpful comments on the essay. I wish to thank Murray Gell-Mann, Stephen Jay Gould, Daniel C. Dennett, Russell Jacoby, Stewart Brand, and David Shipley.

I am grateful to Judy Herrick, who for the past three years has presented me with thousands of pages of accurate transcriptions. I also want to thank my line editor, Sara Lippincott, for her time, effort, diligence, and valuable suggestions. And to Bob Asahina, my editor at Simon & Schuster, I express my appreciation for a perceptive reading of the manuscript and for his friendship.

A number of friends have read and commented on drafts of the manuscript. I would like to thank Wim Coleman, Pat Perrin, Clifford Stoll, Howard Rheingold, Stewart Brand, and Kevin Kelly.

Finally, special thanks and appreciation to Katinka Matson and our son, Max Brockman, for their patience and support.

INTRODUCTION

THE EMERGING THIRD CULTURE

The third culture consists of those scientists and other thinkers in the empirical world who, through their work and expository writing, are taking the place of the traditional intellectual in rendering visible the deeper meanings of our lives, redefining who and what we are.

In the past few years, the playing field of American intellectual life has shifted, and the traditional intellectual has become increasingly marginalized. A 1950s education in Freud, Marx, and modernism is not a sufficient qualification for a thinking person in the 1990s. Indeed, the traditional American intellectuals are, in a sense, increasingly reactionary, and quite often proudly (and perversely) ignorant of many of the truly significant intellectual accomplishments of our time. Their culture, which dismisses science, is often nonempirical. It uses its own jargon and washes its own laundry. It is chiefly characterized by comment on comments, the swelling spiral of commentary eventually reaching the point where the real world gets lost.

In 1959 C.P. Snow published a book titled *The Two Cultures*. On the one hand, there were the literary intellectuals; on the other, the scientists. He noted with incredulity that during the 1930s the literary intellectuals, while no one was looking, took to referring to themselves as "the intellectuals," as though there were no others. This new definition by the "men of letters" excluded scientists such as the astronomer Edwin Hubble, the mathematician John von Neumann, the cyberneticist Norbert Wiener, and the physicists Albert Einstein, Niels Bohr, and Werner Heisenberg.

How did the literary intellectuals get away with it? First, people in the sciences did not make an effective case for the implications of their work. Second, while many eminent scientists, notably Arthur Eddington and James Jeans, also wrote books for a general audience, their works were ignored by the self-proclaimed intellectuals, and the value and importance of the ideas presented remained invisible as an intellectual activity, because science was not a subject for the reigning journals and magazines.

In a second edition of *The Two Cultures*, published in 1963, Snow added a new essay, "The Two Cultures: A Second Look," in which he optimistically suggested that a new culture, a "third culture," would emerge and close the communications gap between the literary intellectuals and the scientists. In Snow's third culture, the literary intellectuals would be on speaking terms with the scientists. Although I borrow Snow's phrase, it does not describe the third culture he predicted. Literary intellectuals are not communicating with scientists. Scientists are communicating directly with the general public. Traditional intellectual media played a vertical game: journalists wrote up and professors wrote down. Today, third-culture thinkers tend to avoid the middleman and endeavor to express their deepest thoughts in a manner accessible to the intelligent reading public.

The recent publishing successes of serious science books have surprised only the old-style intellectuals. Their view is that these books are anomalies—that they are bought but not read. I disagree. The emergence of this third-culture activity is evidence that many people have a great intellectual hunger for new and important ideas and are willing to make the effort to educate themselves.

The wide appeal of the third-culture thinkers is not due solely to their writing ability; what traditionally has been called "science" has today become "public culture." Stewart Brand writes that "Science is the only news. When you scan through a newspaper or magazine, all the human interest stuff is the same old he-said-she-said, the politics and economics the same sorry cyclic dramas, the fashions a pathetic illusion of newness, and even the technology is predictable if you know the science. Human nature doesn't change much; science does, and the change accrues, altering the world irreversibly." We now live in a world in which the rate of change is the biggest change. Science has thus become a big story.

Scientific topics receiving prominent play in newspapers and magazines over the past several years include molecular biology, artificial intelligence, artificial life, chaos theory, massive parallelism, neural nets, the inflationary universe, fractals, complex adaptive systems, superstrings, biodiversity, nanotechnology, the human genome, expert systems, punctuated equilibrium, cellular automata, fuzzy logic, space biospheres, the Gaia hypothesis, virtual reality, cyberspace, and teraflop machines. Among others. There is no canon or accredited list of acceptable ideas. The strength of the third culture is precisely that it can tolerate disagreements about which ideas are to be taken seriously. Unlike previous intellectual pursuits, the achievements of the third culture are not the marginal disputes of a quarrelsome mandarin class: they will affect the lives of everybody on the planet.

The role of the intellectual includes communicating. Intellectuals are not just people who know things but people who shape the thoughts of their generation. An intellectual is a synthesizer, a publicist, a communicator. In his 1987 book *The Last Intellectuals*, the cultural historian Russell Jacoby bemoaned the passing of a generation of public thinkers and their replacement by bloodless academicians. He was right, but also wrong. The third-culture thinkers are the new public intellectuals.

America now is the intellectual seedbed for Europe and Asia. This trend started with the prewar emigration of Albert Einstein and other European scientists and was further fueled by the post-Sputnik boom in scientific education in our universities. The emergence of the third culture introduces new modes of intellectual discourse and reaffirms the preeminence of America in the realm of important ideas. Throughout history, intellectual life has been marked by the fact that only a small number of people have done the serious thinking for everybody else. What we are witnessing is a passing of the torch from one group of thinkers, the traditional literary intellectuals, to a new group, the intellectuals of the emerging third culture.

• • •

Who are the third-culture intellectuals? The list includes the individuals featured in this book, whose work and ideas give meaning to the term: the physicists Paul Davies, J. Doyne Farmer, Murray Gell-Mann, Alan Guth, Roger Penrose, Martin Rees, and Lee Smolin; the evolutionary biologists Richard Dawkins, Niles Eldredge, Stephen Jay Gould, Steve Jones, and George C. Williams; the philosopher Daniel C. Dennett; the biologists Brian Goodwin, Stuart Kauffman, Lynn Margulis, and Francisco J. Varela; the computer scientists W. Daniel Hillis, Christopher G. Langton, Marvin Minsky, and Roger Schank; the psychologists Nicholas Humphrey and Steven Pinker.

During the past three years, I have had ongoing one-on-one discussions with the above-mentioned scientists about their own work and the work of other scientists included in the book. The result is not an anthology, nor is it an overview. I see it as an oral history of a dynamical emergent system, a celebration of the ideas of third-culture thinkers who are defining the interesting and important questions of our times. Here they are communicating their thoughts to the public and to one another. It is an exhibition of this new community of intellectuals in action.

The selection of scientists included in this book is, obviously, far from comprehensive. Many important contributors to the third culture, including social, behavioral, and anthropological scientists, are not here. In addition, the contributions of science journalists—many of whom are distinguished writers and notable thinkers—must also be recognized; their books have provided the public with a wider understanding and greater appreciation of the work and ideas identified with the third culture.

Some of the scientists in the book I work with professionally: they are clients of my literary agency; others are not. (Indeed, the great percentage of scientists I represent are not included here.) The selection is serendipitous, and has to do with my personal scientific interests as well as with the availability of the scientists themselves. The ideas presented are speculative; they represent the frontiers of knowledge in the areas of evolutionary biology, genetics, computer science, neurophysiology, psychology, and physics. Some of the fundamental questions posed are: Where did the universe come from? Where did life come from? Where did the mind come from? Emerging out of the third culture is a new natural philosophy, founded on the realization of the import of complexity, of evolution. Very complex systems—whether organisms, brains, the biosphere, or the universe itself—were not constructed by design; all have evolved. There is a new set of metaphors to describe ourselves, our minds, the universe, and all of the things we know in it, and it is the intellectuals with these new ideas and images—those scientists doing things and writing their own books—who drive our times.

I have taken the editorial license to create a written narrative from my tapes, but although the participants have read, and in some cases edited, the transcriptions of their spoken words, there is no intention that the following chapters in any way represent their writing. For that, read their own books. I have also made the assumption that the views of scientists such as Richard Dawkins and Martin Rees on natural selection and cosmology are of more interest to readers than my own ideas on such subjects. I have thus written myself (and my questions) out of the text. Finally, remarks made about other scientists and their work are general in nature and were not made as responses to the text.

•   •   •

STEPHEN JAY GOULD: The third culture is a very powerful idea. There's something of a conspiracy among literary intellectuals to think they own the intellectual landscape and the reviewing sources, when in fact there are a group of nonfiction writers, largely from the sciences, who have a whole host of fascinating ideas that people want to read about. And some of us are decent writers and express ourselves well enough.

The British Nobelist Peter Medawar, a very humanistically and classically educated scientist, said it was unfair that a scientist who didn't know art and music pretty well was, among literary people, considered a dolt and a philistine, whereas literary people don't think they need to know any science in order to be considered educated; all an educated person has got to know is art and music and literature, but not any science.

That just isn't right, and it doesn't reflect reality, either. It may be that of the two hundred and eighty million people in America, not a very high percentage understands science well, but among people who buy books—which may not be a high percentage of the American population but is a high absolute number—interest is very strong.

MURRAY GELL-MANN: Scientists used to write books for the interested public—those people who care about science and have a certain amount of scientific literacy. There was a time when that activity nearly died out, at least in this country. It's a very healthy trend that we are now seeing, with serious scientists once again writing about their work, dealing with the public directly as well as through journalist intermediaries. Some scientists have always been better than others at writing general material, and some are broader than others in their culture. But among scientists who have done interesting work, there have always been and will always be a number who can communicate quite successfully with the public and don't need to depend on intermediaries.

Unfortunately, there are people in the arts and humanities—conceivably, even some in the social sciences—who are proud of knowing very little about science and technology, or about mathematics. The opposite phenomenon is very rare. You may occasionally find a scientist who is ignorant of Shakespeare, but you will never find a scientist who is *proud* of being ignorant of Shakespeare.

DANIEL C. DENNETT: The hallmark of the recent successes among science books is related to the interdisciplinary nature of many of the new scientific endeavors. Professors are writing for colleagues in other disciplines. Thus, they must write in plain English and avoid the jargon of their own field. If I were writing a book just for philosophers—my own field—I would write it that way, and for the same reason. I know this jargon problem is there in every discipline, but it's there in spades in philosophy. A lot of the bad artifactual problems that arise in philosophy arise from experts talking to experts. The worst sin an expert can commit when talking to another expert is to overexplain, to talk down—this is insulting. So experts always err on the side of underexplaining. As a result, they tend to talk past each other. They don't realize that they aren't sharing common assumptions. Then you get these tremendous edifices of conflict, which are based on rather simple fundamental misunderstandings on a low level.

There's a profound difference between the Anglophone university tradition and the European. In Europe, professors profess. They have their podium, and boy oh boy, they lay it out, and you take notes and you don't ask questions; there's a certain cachet in being hard to understand and being inaccessible. This is the way you make your reputation, by being obscure. That doesn't happen to anyone in the English-speaking university tradition, to the same degree; I don't know whether that has much to do with science. But you can see it also influencing the nonscientific or semiscientific or philosophical writing of continental scientists. Jacques Monod and François Jacob are two examples. They aspired to be philosophers—which is fine, so do lots of Anglophone scientists—but they aspired to be continental philosophers, which led them into some deeper, darker waters than they knew how to swim in.

RICHARD DAWKINS: I do feel somewhat paranoid about what I think of as a hijacking by literary people of the intellectual media. It's not just the word "intellectual." I noticed, the other day, an article by a literary critic called "Theory: What Is It?" Would you believe it? "Theory" turned out to mean "theory in literary criticism." This wasn't in a journal of literary criticism; this was in some general publication, like a Sunday newspaper. The very word "theory" has been hijacked for some extremely narrow parochial literary purpose—as though Einstein didn't have theories; as though Darwin didn't have theories.

I applaud the idea that scientists, and scholars generally, can communicate their original ideas to one another in books that are read by people in other fields. My own books have been both popularizations of material already familiar to scientists and original contributions to the field which have changed the way scientists think, albeit they haven't appeared in scientific journals or been languaged up with incomprehensible jargon. They've been written in terms that any intelligent person can understand. I should like to see more people doing that.

P.B. Medawar said that there are some fields that are genuinely difficult, where if you want to communicate you have to work really hard to make the language simple, and there are other fields that are fundamentally very easy, where if you want to impress people you have to make the language more difficult than it needs to be. And there are some fields in which—to use Medawar's lovely phrase—people suffer from "physics envy." They want their subject to be treated as profoundly difficult, even when it isn't. Physics genuinely is difficult, so there's a great industry for taking the difficult ideas of physics and making them simpler for people to understand; but, conversely, there's another industry for taking subjects that really have no substance at all and pretending they do—dressing them up in a language that's incomprehensible for the very sake of incomprehensibility, in order to make them seem profound.

STEVE JONES: The best way of assessing the "third culture" idea is to ask, "Has there ever been more than one culture?" That's the central question. Is learning divisible, or is it seamless? From 1550 to around 1950 the answer was obvious: culture is culture—although, after Milton, nobody could know everything. Then C.P. Snow came up with a Christmas-cracker motto describing a division that may or may not have been there. I'm not convinced that he overturned four hundred years of civilization, although he may have punctured the egos of a few of the arrogant literary mediocrities who surrounded him.

The question now, as in Snow's day, is whether there's a culture to which every educated person can cleave. The answer is that if there isn't, there certainly ought to be. If you aren't someone who can talk in general terms about scientific as well as nonscientific issues, you aren't civilized.

PAUL DAVIES: It's difficult to disentangle the problem of the two cultures and the third culture from the class and regional prejudices that pervade British society. One of the distinctive features of British intellectual life is its dominance by just two universities: Oxford and Cambridge. Most of the politicians and members of the establishment—the civil service, the media, and the people who control the media—are Oxford arts graduates. As a result, the public's perception of an intellectual is a graying, bespectacled gentleman who studies Greek mythology, drinks sherry, and punts leisurely and contemplatively on the river through the grounds of an ancient college. And with this perception is accorded a status suggesting that it's the arts and literary intellectuals who have a God-given monopoly on the great issues of existence.

It's only in recent years that scientists have exercised any sort of influence over what we might call the big questions, and this influence has created a very ugly backlash. The fact that scientists are starting to be heard, capturing not only the minds but the hearts of the population—as evidenced by the phenomenal success of science books—is provoking what seems to be a territorial squeal from the literary side. The backlash has taken the form of hysterical ranting in newspapers and periodicals, and a spate of books denouncing scientists as arrogant and self-serving frauds.

Few intellectuals in Britain make any attempt to understand science, and clearly feel out of their depth with the issues being presented in recent books such as Stephen Hawking's *A Brief History of Time.* Some of the backlash seems to stem from a sense of helplessness in the face of this ignorance. "I'm well-educated," they say, "and I can't make sense of this. Therefore it must be bunk!" A few years ago, when the discovery of ripples in the cosmic microwave background radiation was announced, the influential and noted journalist Bernard Levin basically trashed the entire cosmological program as being unworthy of serious comment. He said, for example, that the big-bang theory didn't have a shred of evidence going for it. This is a grossly misleading statement, because, of course, there's a lot of evidence for it. Another journalist who has made scientists a target is Brian Appleyard. In the Foreword of his best-selling book *Understanding the Present,* he says he was moved to write it because of the outrage he felt after interviewing Hawking. He was upset by what he saw as the arrogance of scientists attempting to pronounce on deep issues of God, existence, and humanity. You get the impression that this kind of response—to important and exciting scientific discoveries that change the way that we look at the world—is a sort of knee-jerk territorial reaction. For years and years scientists were ignored because they were not heard; now that they're starting to be heard, they're being stamped on by an intellectual mafia.

NICHOLAS HUMPHREY: There's terror among the British intelligentsia that culture has passed them by. They went to school, learned their classics, learned their English literature, thought of scientists as some kind of nerds. What went on in the chemistry or the biology labs was beneath contempt for these intellectuals who were in touch with Plato and Aristotle and Julius Caesar. Such people, who are used to being dominant in our culture, are suddenly scared. Since they don't understand science, their only defense is to say that it doesn't matter. But they're fighting a losing battle. People are voting with their feet. Who listens to what nowadays? Who watches what on TV? Who's buying what books?

W. DANIEL HILLIS: The scientists who are representative of an emerging third culture are not typical scientists but those who in some sense have participated in the wider world—people who have discovered that the problems they're working on don't fit within the neat structures of their internal disciplines. Many of the scientists who write popular books do so because there are certain kinds of ideas that have absolutely no way of getting published within the scientific community. There's a tradition for this. A hundred years ago, the intellectuals were the scientists—natural philosophers.

Something that *is* new is that people are compelled to see that science is relevant; it's changing their lives, much faster than they want it to. For a while, people were content to let the scientists do science, and trusted them to understand that kind of stuff: it was all so abstract. Now there are people who realize that their lives are completely different because of a bunch of things they don't understand.

We're going through a qualitative change. People no longer have a view of the future stretching out even through their own lifetimes, much less through the lifetimes of their children. They realize that things are moving so fast that you can't really imagine the life your child is going to lead. That's never been true before, and it's clear that the course of that change and that discontinuity is science, somehow. Anybody who is not brain dead wants to try to get ahold of things—is strongly motivated to do so—and one way to do it is to read books by scientists.

A problem the third culture faces is that scientists often look down on other scientists when they explain their ideas clearly to nonscientists. When you get somebody who's very articulate, like Gould or Dawkins, other scientists get a little bit jealous, because those two are explaining to the public the issues we're arguing about. That's particularly true in biology. There's a feeling in biology that scientists should keep their dirty laundry hidden, because the religious right are always looking for any argument between evolutionists as support for their creationist theories. There's a strong school of thought in biology that one should never question Darwin in public. But it's also true that "popularizer" is a pejorative term among

scientists generally. A popularizer is somebody who explains what
the issues are in ways that people can understand. I think it's ridiculous that scientists don't respect such people. In any other field, explaining to a congressional committee why what you're doing is
exciting and wonderful would be considered a service to the field. In
science, you're treated almost like somebody who has betrayed the
secret club.

ROGER SCHANK: I'm on the editorial board of the *Encyclopedia Britannica*, and one of the things that went on a year or two ago was
this discussion of who was going to be taking care of the encyclopedia in the future, and what would be in it. The board, who are all
these literary types, decided it would let computer people in, because
the world was getting to be computerized. And Clifton Fadiman said
that he supposed we'd have to resign ourselves to the fact that minds
less educated than ours would soon be in charge of *Encyclopedia Britannica*. I said, "Hey! How did you decide that I'm less educated
than you are?" And he actually got out of it—he said, "Oh, I didn't
mean you! You're a very phenomenal and unusual computer scientist."

But I'm not a phenomenal and unusual computer scientist at all.
What's interesting about such people in the literary world is that they
somehow think that if you don't know the classics you're unedu-
cated, whereas it's O.K. for them not to know beans about science.
And I don't understand why that's O.K.

We're living in a world in which no one can be an expert on everything; there's too much to know. So the idea of being very broad is no
longer an appropriate model—everyone's going to have limitations.
Somehow, we've set out these limitations. The ultimate one—the one
society cannot put up with—is that you don't know the classics. Mor-
timer Adler, the head of the *Britannica* editorial board, says the
same thing. We've argued a lot about the "great books." He's had a list
of the great books printed; they're very interesting books, but the fact
of the matter is that they leave out almost all of what we've learned
in the last hundred years.

I've been reading a lot lately about consciousness. I'm interested
in this subject now, and I want to find out as much as I can about it.
And finding these things, written by many different authors, has been
easy for me because of an index Adler has put together called *The*

Syntopicon. I've been able to find remarks on the subject by Thomas Aquinas and Montaigne and Aristotle—the authors Adler has listed under "consciousness." These people have a vague hand-waving notion of what consciousness is about, with a religious tinge to it. Their work wouldn't fly at all in modern academics. Yet we're being told that if you haven't read them you aren't educated. Well, I'm reading them, but I'm not learning much from them. What I'm learning is that people have struggled with these ideas for the last two thousand years and haven't been all that clever about it a lot of the time. Now, with the computer metaphor, and a different way of looking at the idea of consciousness, we have entirely different and new and interesting things to say, and yet the Clifton Fadimans of the world wouldn't read what we have to say. I'm willing to bet he didn't read Dan Dennett's *Consciousness Explained*, for example—but it's O.K., he's still educated.

We got pushed out of the intellectual circle, for reasons that aren't interesting. Maybe that's why scientists are writing popular books: because they're some of the most interesting people in society and they're not considered great intellectuals. But then maybe neither are the literary people, right now; I'm not sure this is a country that admires intellectuals much.

J. DOYNE FARMER: One of the biggest problems for society in general is synthesizing knowledge. Society is a very complex organism, and the need for increasing specialization has driven everyone to levels of specialization that have created enormous information barriers. Newton published in the *Philosophical Transactions of the Royal Society*, and up through the nineteenth century physicists were still publishing in journals that had titles with "Philosophy" in them, and there wasn't a clear distinction drawn. They were natural philosophers. Increasingly in the twentieth century, science has become more and more separated.

There was a wave of physicists who emerged in the 1950s— Richard Feynman being a prime example—who disdained philosophy and thought it wasn't something a physicist should do. In a certain sense, this attitude arose with good reason. When you look at the direction philosophy took in the twentieth century—it's pretty dismal stuff.

But things were very different for Einstein and Bohr and people in that generation. The physicists who made the big breakthroughs in the 1920s were, by and large, well educated in philosophy. Einstein, for example, quotes Kant frequently, and viewed philosophical education as something that was important for a physicist to have. In fact, many physicists at the time wrote philosophical papers, and the connection was still there. By the 1950s it was completely lost, and my generation grew up hearing not only that this isn't something you should spend your time doing but that you could get into serious trouble for being a philosopher. If you wrote a paper in a philosophy journal—or worse, if you wrote a popular book—you were endangering your reputation.

MARTIN REES: Most of those with editorial control in the media have a primarily literary education and are now increasingly untypical, in background and interests, of intelligent readers in general. This problem is, incidentally, even worse in the U.K., because our education system is more specialized, and many people who go on to university had no exposure to any scientific subject after the age of fifteen.

There's an awareness that there are general concepts, like chaos, that can be quantified and applied to a lot of unrelated contexts. This awareness is having a very good effect: it brings together people who might otherwise have languished in separate disciplines. There's obviously a gap between those who are at ease with mathematics and those who are not. This is a big problem for all of us who try to explain physical ideas to a general readership. There's clearly a demand for this, and most of those who control the media perhaps don't appreciate the fact that more than half the readers of the quality press must be people with some scientific training, and that there's a demand for fairly sophisticated—although not too mathematical—discussion of general issues.

LEE SMOLIN: In addition to having a theory of quantum gravity, I have the need to communicate it outside the physics community. When I listen to people in the humanities, I realize that they have similar problems with regard to communicating difficult ideas. I can't read them line by line, because the language is based on Hegel and Heidegger, or whomever, and it doesn't make any sense to me. They have some romantic idea about being difficult, and this is wrong. Why they do it, or why it's popular, is something I don't understand. I don't want to push it too much, because it's quite enough to ask this question inside science.

I am not incomprehensible. Given an hour or so, I can make myself comprehensible. One of the differences between the traditions of science and the humanities is that the humanities have become traditions of reading and writing. People in these fields don't talk to each other. They sit at home and they sit in their offices and they construct sentences and paragraphs, and they don't speak to each other. Scientists speak to each other, first and foremost. Our culture is verbal, and we know how to talk to people. Go to a talk given by somebody in philosophy or literary theory. Notice that they invariably will read something that they've written, word for word. Very few scientists will ever do that.

For me, the scientists grouped under the name of the third culture represent more than just a set of academics who write and speak to the general public. There are philosophical ideas that they share, to a greater or lesser extent. If I may be very optimistic, I see a kind of rebirth of the tradition of natural philosophy, but based on a new picture of the world—a picture different from the one that the original, seventeenth-century natural philosophers shared. This new spirit has several overarching themes, which are not hard to state. Of first importance is the idea that the world is not static or eternal, it evolves in time. The world was different in the past and it will be different in the future. In the nineteenth century, we discovered that this was true of the biological world, and in the twentieth century we've discovered that it's true of the universe as a whole. In my opinion, we're only now beginning to realize the implications of these discoveries, just as it took more than a century for the implications of Copernicus's discoveries to become evident.

The second theme is that we're beginning to realize not only that it's unnecessary to think in terms of an intelligent designer but that the idea that the complexity and beauty we see around us was intended by a single intelligence is silly. Instead, we understand, in the biological context, that the living world has created itself—organized itself—because of the action of simple principles, primarily natural selection, that inevitably operate. I believe that the same will turn out to be true about the laws of physics and the structure of the cosmos.

The third theme is complexity: that the fact that the world is complex is essential and not accidental, that there's an enormous variety of things and phenomena in the world. Finally, in such a complex self-organized world, all properties of things are relational. The notion of absolute properties—of, say, biological species—has become as obsolete as Newton's conception of absolute space and time.

I sometimes see these themes also in the work of artists, such as Saint Clair Cemin and Donna Moylin. Of course, there are many artists—and many "intellectuals" who write about art—who are still caught in the trap of Nietzsche, playing with death and violence and negativity, playing out the death of some old and obsolete notions of the world. But these people are more and more irrelevant; what's interesting is that some artists have understood that the world's not going to end soon, that the twenty-first century is going to be an extraordinary time, and that the time is now to begin imagining what direction the human community may go in.

Part One

THE EVOLUTIONARY

IDEA

The universe is changing in time, and it has evolved from something simpler to something more complex. That is the lesson to be learned from recent advances in evolutionary theory; the emergence of order has colored biology since Darwin and twentieth-century cosmology alike.

In Darwin's day, the exact manner of the inheritance of characteristics was not known; Darwin himself believed that certain characteristics were acquired by an organism as a result of environmental change and could be passed to the organism's offspring, an idea popularized by the French naturalist Jean-Baptiste Lamarck. In 1900, the work done by Mendel some fifty years earlier was brought to light, and the gene, though its exact nature was unknown at the time, became a player in "the modern synthesis" of Mendel and Darwin. This synthesis, which reconciled genetics per se with Darwin's vision of natural selection, was carried out in the early 1930s by R.A. Fisher, J.B.S. Haldane, and Sewall Wright, and augmented a few years later by the work of the paleontologist George Gaylord Simpson, the biologist Ernst Mayr, and the geneticist Theodosius Dobzhansky, who expanded on this neo-Darwinian paradigm. Nevertheless, there is still discord in the ranks of evolutionary biologists. The principal debates are concerned with the mechanism of speciation; whether natural selection operates at the level of the gene, the organism, or the species, or all three; and also with the relative importance of other factors, such as natural catastrophes.

Among the evolutionary biologists, George C. Williams is the senior figure in the book. People outside the field who take an interest in evolution are much more likely to think of Stephen Jay Gould or the British evolutionary biologist Richard Dawkins; few laypeople have heard of Williams. Yet nearly all evolutionary biologists, even those who do not agree with him, admire him. Williams was the first to emphasize that it was the gene on which natural selection acted. In this regard, he precedes Richard Dawkins, with whom he shares a great many ideas, and he is in a different camp from Stephen Jay Gould, who has a hierarchical theory of selection processes, of which the gene is only one level. Williams' book *Adaptation and Natural Selection,* published in 1966, was a treatise on what has become known as ultra-Darwinism. In recent work, Williams describes the gene as having a "codical" as well as a physical character—that is, he views the gene as a package of information, not an object.

Stephen Jay Gould is known among evolutionary biologists for three things: "punctuated equilibria," a theory he developed with Niles Eldredge which says that one species does not gradually turn into another but that species emerge suddenly; resuscitation of the study of relationships between embryology and evolution, as expressed in his landmark book *Ontology and Phylogeny*; and a famous paper, written with the population geneticist Richard Lewontin, titled "The Spandrels of San Marco and the Panglossian Paradigm; A Critique of the Adaptationist Program." He is also widely read. His books *(Wonderful Life, Bully for Brontosaurus)* are on best-seller lists around the world.

Paleontologists study the fossil record. The questions they face concern such things as long-term patterns in the history of life, and the extinction of species over millions of years. In this regard, Gould carries forward the attitude of many paleontologists towards evolutionary biology: namely, skepticism regarding the domains and the powers of natural selection. He is often critical of the ultra-Darwinian views of mainstream evolutionists such as Williams and Dawkins.

Gould is attempting to weave together three themes in order to expand the current Darwinian model: the first is the hierarchical theory of natural selection; the second notes the limitations on evolution imposed by biological constraints on adaptation; the third concerns catastrophic events in geologic time, which can cause mass extinctions. Because Gould is a prolific and gifted popularizer, the educated public, at least in America, assumes that his approach to evolutionary biology is the mainstream position. Not so; he is a critic of the mainstream, which is dominated by Williams, the English evolutionary biologist John Maynard Smith, and Dawkins.

Richard Dawkins is considered by his peers to be the ultimate ultra-Darwinist. He is also a gifted writer, who is known for his popularization of Darwinian ideas as well as for original thinking on evolutionary theory. He has invented telling metaphors that illuminate the Darwinian debate: His book *The Selfish Gene* argues that genes—molecules of DNA—are the fundamental units of natural selection, the "replicators." Organisms, including ourselves, are "vehicles," the packaging for "replicators." The success or failure of replicators is based on their ability to build successful vehicles. There is a complementarity in the relationship: vehicles propagate their replicators, not themselves; replicators make vehicles. In *The Extended Phenotype*, he goes beyond the body to the family, the social group, the architecture, the environment that animals create, and sees these as part of the phenotype—the embodiment of the genes. He also takes a Darwinian view of culture, exemplified in his invention of the "meme," the unit of cultural inheritance; memes are essentially ideas, and they, too, are operated on by natural selection.

Brian Goodwin looks on biology as an exact science, and sees the "new biology" less as a historical science than as an enterprise similar to physics in its emphasis on principles of order. He represents the structuralist approach, which resonates with the Scottish zoologist D'Arcy Thompson's idea that evolutionary variation is constrained by structural laws; not all forms are possible. Goodwin is passionately opposed to the reductionist view of the ultra-Darwinians, and much more comfortable with the complexity ideas of Stuart Kauffman and with Francisco Varela's holistic approach to biology.

Steve Jones is a highly regarded geneticist and snail biologist. He is interested in why so much diversity exists in animals and plants: why no two individuals are alike. Surely, it can be argued, natural selection should instead inevitably lead to the evolution of one perfect form for each species. He works on the striking variety of shell color and banding patterns in the land snail *Cepaea nemoralis. Cepaea* has been seen as an archetype of diversity since the nineteenth century. In the 1950s, the English biologists Arthur Cain and Phillip Sheppard argued that such apparently trivial differences were under the action of natural selection (in this case because birds would attack the conspicuous forms). Jones finds that climate is also involved and—most important—that differences in microclimate on the scale of a few inches can alter the behavior and survival of snails of different pattern. Ecologically complex habitats hence foster genetic diversity. Jones has been writing and lecturing about science to a general audience for ten years. His most recent book, *The Language of the Genes,* won the 1994 Science Book Prize.

Niles Eldredge, a paleontologist who is closely associated with Stephen Jay Gould, believes that species are information repositories; the theory of punctuated equilibria provides evidence that species, once they originate, tend to remain stable, while change occurs by rapid events of branching by small populations splitting off from ancestral species. Eldredge is known for his work on the hierarchical structure of biological systems and for his criticisms of the adaptationist program. Since the late 1960s, he and Gould, the "naturalists," have been locked in intellectual debate with George Williams, John Maynard Smith, and Richard Dawkins over such issues as where selection takes place. Eldredge believes that metaphors of competition, while appropriate at the level of reproductive biology, fail as an explanatory principle for large-scale biological systems. He also makes the point that while the ultra-Darwinians consider themselves reductionists, they go down only to the genes-within-populations level, and don't pay much attention to molecular biology.

Biologist Lynn Margulis would extend evolutionary studies nearly four billion years back in time. Her major work has been in cell evolution, in which the great event was the appearance of the eukaryotic, or nucleated, cell—the cell upon which all larger life-forms are based. Nearly thirty years ago, she argued for its symbiotic origin: that it arose by associations of different kinds of bacteria. Her ideas were generally either ignored or ridiculed when she first proposed them; symbiosis in cell evolution is now considered one of the great scientific breakthroughs.

Margulis is also a champion of the Gaia hypothesis, an idea developed in the 1970s by the free-lance British atmospheric chemist James E. Lovelock. The Gaia hypothesis states that the atmosphere and surface sediments of the planet Earth form a self-regulating physiological system—Earth's surface is alive. The strong version of the hypothesis, which has been widely criticized by the biological establishment, holds that the earth itself is a self-regulating organism; Margulis subscribes to a weaker version, seeing the planet as an integrated self-regulating ecosystem. She has been criticized for succumbing to what George Williams calls the "God-is-good" syndrome, as evidenced by her adoption of metaphors of symbiosis in nature. She is, in turn, an outspoken critic of mainstream evolutionary biologists for what she sees as a failure to adequately consider the importance of chemistry and microbiology in evolution.

Chapter 1

GEORGE C. WILLIAMS
"A Package of Information"

NILES ELDREDGE: I remember the English evolutionary geneticist John Maynard Smith remarking to me that he was astonished to find out that George Williams wasn't in our National Academy. Williams finally got elected in 1993. When I visited him in Stony Brook in the mid-1980s, he told me he was having a hard time getting grant support for his research, and I couldn't believe that. The two thoughts converged, because George really is the most important thinker in evolutionary biology in the United States since the 1959 Darwin centennial. It's astonishing that he hasn't gotten more credit and acclaim. He's a shy guy, but a very nice guy, and a very deep and a very careful thinker. I admire him tremendously, even though we've been arguing back and forth for years now.

• • •

GEORGE C. WILLIAMS is an evolutionary biologist; professor emeritus of ecology and evolution at the State University of New York at Stony Brook; author of Adaptation and Natural Selection: A Critique of Some Current Evolutionary Thought *(1966),* Sex and Evolution *(1975),* Natural Selection: Domains, Levels, and Challenges *(1992), and coauthor (with Randolph Nesse, M.D.) of* Why We Get Sick *(1995).*

GEORGE C. WILLIAMS: Evolution, in the sense of long-term change in a sexually reproducing population, depends on the relative rates of survival of competing genes. Given that organisms may find themselves in an environment where there are close genealogical relatives, it follows that an organism is expected to react to cues of kinship in a certain way, so as to discriminate among the individuals it encounters on the basis of kinship, and be more benign and cooperative toward closer kin than more distant kin or nonrelatives.

My interest in evolution started in the summer of 1947, when I spent six weeks in the Painted Desert with a paleontologist named Sam Welles, who had a group of students there, officially in a summer course, but we spent most of the time swinging picks and shovels, digging fossils, as part of Welles' research project. He was a specialist in Triassic amphibians. Evenings were spent sitting around the campfire talking about things like evolution. For the first time in my life, people—real biologists, real scholars—were willing to sit and listen to my opinions. I was twenty-one years old. I certainly became interested in many aspects of evolution then, and shortly after that I signed up at the University of California at Berkeley for a course in evolution with Ledyard Stebbins, who at the time, and for quite a while thereafter, was the world's primary expert in evolution with respect to things botanical. Stebbins' course introduced me to Theodosius Dobzhansky's *Genetics and The Origin of Species*. Stebbins was great, but Dobzhansky's book was what got me interested in natural selection as a process.

At the University of Chicago, my job was strictly teaching. I was in their early-entrant undergraduate program—taught freshmen and sophomores biology. They had a great-books approach. We read Darwin, Mendel, and others. Also I attended seminars by people such as Alfred Emerson, the termite specialist and recognized authority on things evolutionary. I found his ideas absolutely unacceptable. That motivated me to do something. If it was biology Emerson was discussing, I would be better off selling insurance.

I remember especially his lecture on the role of death in evolution. He was all in favor of death, and said that the reason we grow old and die is to make room for successors, so that they can have a chance. This seemed so totally impossible, given that evolution proceeds by natural selection. There was absolutely no logical way you could reconcile his ideas with Darwinism, even though he claimed to be a Darwinist.

This initiated my first theoretical obsession: the evolution of senescence—the decline in adaptive performance with age. You can't run as fast at sixty as you could at thirty. On the way home that evening, talking about the problem with my wife, I independently came up with an idea that Peter Medawar is chiefly responsible for and published in 1952, although he may have published something that foreshadowed it in the 1940s—and that is that the effectiveness of selection in maintaining adaptation is essentially the product of reproductive value and survival.

The survival factor is easier to appreciate. If you're more likely to be alive at thirty than at sixty, then selection will be more effective at maintaining adaptation at thirty than at sixty. At an age you'd be extremely unlikely to survive to, such as one hundred years old, adaptation would be a lost cause, and selection wouldn't be concerned with it.

As the effectiveness of selection declines, the effectiveness of its products declines. This explains the rising mortality rate that comes with age. It seemed to me at the time, and still does, that this is an inevitable conclusion, arising from just the simple fact of mortality. If there's any possibility of dying, at any age, then you're less likely to be alive at a later age than you are at an earlier age.

Another one of Alfred Emerson's ideas was that evolution is much more concerned with cooperation than with competition. It seemed to me to be very much the other way around, and that there was something very special about the social insects which accounted for their extreme cooperativeness. That special thing was their kinship—high levels of kinship within the colony. This was the focus of a theoretical paper I published in 1957. It was a model of natural selection between families; now I think that's a silly way to do it, but at the time I wasn't smart enough to think of the kin-selection idea, which was some years later worked out by William D. Hamilton. In extreme models, this kind of selection can lead to things like forgoing reproduction, if in so doing you can, for example, more than double the reproduction of a full sib. The full sib is half as good as you are genetically—that is, from the standpoint of getting your genes into future generations. In the social insects, of course, sisters may have a three-quarter relationship, because if they share a father then all the genes they get from the father are exactly the same.

These early experiences kindled an interest that has never gone away, and resulted in *Adaptation and Natural Selection,* my first book-length publication on this and related matters. By then I had worked on the problem of senescence and on cooperation between relatives, but I had a long list of other problems that interested me.

At that time, group selection was not explicit. V.C. Wynne-Edwards' big book on group selection—*Animal Dispersion in Relation to Social Behaviour*—came out in 1962, but I discovered it only after I was largely finished with *Adaptation and Natural Selection.* I submitted the manuscript in late 1963, and it referred to Wynne-Edwards' work, but I brought it in as a late revision of the manuscript.

There was some group-selection modeling prior to that, and explicit use of group-selection ideas by Alfred Emerson and A.H. Sturtevant, in a paper published in 1938. In 1945, Sewall Wright presented a group-selection model, in a book review of George Simpson's *Tempo and Mode in Evolution.* But the group-selection model wasn't easy to find if you didn't know about it already. Mostly, the group-selection idea was necessary to the way people were thinking about adaptation, although—and I find this extremely strange—they didn't realize it. They kept talking about things being for the good of the species. If it's for the good of something, and it's to arise by natural selection, it has to be produced by the natural selection of those somethings. In other words, one species survives as another one goes extinct. The basis of Wynne-Edwards' work on group selection was that you can't have things that work for the good of the group unless you have selection at the level of groups. What he was doing was looking for selection at the level of local breeding populations, and whether they could be called separate species wasn't particularly relevant.

To most people's satisfaction, Wynne-Edwards has been proved wrong. Not that there's no selection at levels higher than the individual or the family, but simply that his particular formulation isn't likely to be a very strong force in evolution. It's now generally conceded that the phenomena he was explaining by this mode of thought are much better explained by other processes: by selection at lower levels, selection among individuals.

For instance, any reproductive restraint—anytime it looks as if individuals aren't reproducing at the maximum possible rate—is explainable simply on the basis of an individual optimal-resource-allocation model. You don't kill yourself trying to do something today if working at it a little bit more easily will enable you to try again tomorrow. Maybe you don't do it at *all* today, if conditions will be much better tomorrow. This kind of thinking explains the fact, for instance, that birds do not necessarily lay as many eggs in a breeding season as they demonstrably might. The allocation of their resources will be much more effective for reproduction with a lower-level expenditure on eggs, which will enable them later to spend more on feeding the young and later still, next year, having another breeding season.

There's a great conceptual deficiency in my earlier work, one that I shared with just about everybody else who was working at the time. I failed to realize what a tremendous problem the existence and prevalence of sexual reproduction is. I got interested in that in the early seventies, and I published a book in 1975 titled *Sex and Evolution*. There are a lot of complications that I didn't appreciate at the time, but John Maynard Smith and Bill Hamilton and many others have advanced our understanding tremendously in the last twenty years.

Richard Dawkins went in the right direction when he made the distinction between replicators and vehicles. David Hull's substitution of the term "interactor" for "vehicle" is a good idea, but that's a minor terminological matter. Dawkins didn't go nearly far enough in making that distinction, because he defines a replicator in a way that makes it a physical entity duplicating itself in a reproductive process. This is fine, but the important distinction lies at a still more basic level. He was misled by the fact that genes are always identified with DNA.

Evolutionary biologists have failed to realize that they work with two more or less incommensurable domains: that of information and that of matter. I address this problem in my 1992 book, *Natural Selection: Domains, Levels, and Challenges*. These two domains will never be brought together in any kind of the sense usually implied by the term "reductionism." You can speak of galaxies and particles of dust in the same terms, because they both have mass and charge and length and width. You can't do that with information and matter. Information doesn't have mass or charge or length in millimeters. Likewise, matter doesn't have bytes. You can't measure so much gold in so many bytes. It doesn't have redundancy, or fidelity, or any of the other descriptors we apply to information. This dearth of shared descriptors makes matter and information two separate domains of existence, which have to be discussed separately, in their own terms.

The gene is a package of information, not an object. The pattern of base pairs in a DNA molecule specifies the gene. But the DNA molecule is the medium, it's not the message. Maintaining this distinction between the medium and the message is absolutely indispensable to clarity of thought about evolution.

Just the fact that fifteen years ago I started using a computer may have had something to do with my ideas here. The constant process of transferring information from one physical medium to another and then being able to recover that same information in the original medium brings home the separability of information and matter. In biology, when you're talking about things like genes and genotypes and gene pools, you're talking about information, not physical objective reality. They're patterns.

I was also influenced by Dawkins' "meme" concept, which refers to cultural information that influences people's behavior. Memes, unlike genes, don't have a single, archival kind of medium. Consider the book *Don Quixote*: a stack of paper with ink marks on the pages, but you could put it on a CD or a tape and turn it into sound waves for blind people. No matter what medium it's in, it's always the same book, the same information. This is true of everything else in the cultural realm. It can be recorded in many different media, but it's the same meme no matter what medium it's recorded in.

In cultural evolution, obviously, the idea of a coffee cup or a table is something that persists. The coffee cups and tables don't persist, they recur as a result of the persistence of the information that tells people how to make coffee cups and tables. It's the same way in biology: hands and feet and noses and so on don't persist, they recur as a result of genetic instructions for making hands and feet and noses. It's the information that lasts and evolves. Obviously, it's because of the physical manifestations of the information that we know about the information. Dawkins has had trouble in convincing people, and this stems from his thinking of the gene as an object—of emphasizing the importance of replication rather than of proliferation of information.

Until you've made the distinction between information and matter, discussions of levels of selection will be muddled. Comparing a gene with an individual, for instance, in discussions of levels of selection, is inappropriate, if by "individual" you mean a material object and by "gene" you mean a package of information. It should be "gene" and "genotype." You have to look at levels of selection in both of these domains, and realize what you're doing. Comparisons of levels of selection should be within the same domain.

Having made the domain distinction, you then go to levels, and you find that in the two domains the levels do not correspond exactly. As a general rule, if we restrict our attention to sexually reproducing populations, there are only two possible levels of selection in the informational—or what I call the codical—domain: the gene and the gene pool. Selection can operate on alternative genes within a population; selection can act on alternative gene pools in a biota. Both of these are evolutionary factors that can produce interesting effects.

In the material domain, on the other hand, selection can operate at the level of alternative individuals, in the usual sense of "individual," or on groups of individuals—such things as insect colonies, or families whether they form elaborate colonies or not. These temporary groupings of individuals give rise to what the biologist David Sloan Wilson calls "trait-group selection," and also to selection between alternative populations. That's the physical basis for selection between gene pools. But the physical levels of selection below that level—for instance, between competing colonies of the same species of social insects—don't have a corresponding level in the codical domain. The events in the competition between insect colonies are recorded at the level of a gene. There are no sufficiently persistent genetic differences among colonies for effective selection in the codical domain. I believe David Wilson agrees with that. He's interested in selection among the interactors in the material domain.

The main messages of my 1966 book are now generally accepted. This would have been the case whether I wrote that book or not. The ideas would have prevailed by today, because people like Hamilton, Dawkins, Robert Trivers, and others were doing work at the same time, more or less, and if there hadn't been a single book in the mid-sixties to deal with the idea of levels of selection, I think one of those people probably would have written it. Dawkins' book *The Selfish Gene* is very much a case in point. It advanced things a lot further than mine did.

My lasting contribution will be for a clarification of the problems of the two domains and the levels of natural selection. I'll also be known as one of the people who first became interested in explaining why there is such a thing as sexual reproduction, and why it's so widespread.

In the future, breakthroughs in evolutionary biology may come in the field of paleontology. Fieldwork now going on will be recognized several decades from now as having provided extremely important information. People I've never heard of are out there digging, looking for pollen grains in lake sediments, or dusting off trilobites from Paleozoic shales. Other important insights will come from people working in traditionally unrelated fields—for instance, on things like conflict between genomes. The most immediately enlightening and convincing work that's going on now is in explanations being advanced for things like genetic imprinting—that is, the fact that in early development the activity of the gene depends upon whether it came from the mother or the father. I'm most involved in a recent publication by the biologist David Haig on genetic conflict in human pregnancy. This may not in fact be the clearest example of genetic imprinting, and certainly it isn't going to be the one most easy to work with, but it's work of this nature that's likely to get people thinking seriously about levels and domains of selection.

My recent work concerns what I call Darwinian medicine—the general applicability of evolutionary ideas in medical research, practice, and education. It arose in conversations with Randolph Nesse, a medical doctor and professor at Ann Arbor. Another important factor for me was a paper by Paul Ewald in 1980.

Ewald started life as an ornithologist and got interested in medicine one day when he got sick. It was an intestinal pathogen that got him—not quite as dramatic as Alfred Russel Wallace getting his inspirations during an attack of malaria. Paul started thinking about the evolutionary interaction between hosts and parasites. That led to his paper on how to use evolutionary ideas to interpret the observations one makes in infectious diseases—the symptoms and signs seen in the host. It struck me that these were extremely important ideas, which should be tremendously useful in medicine.

I had already been thinking about senescence and life histories in general, and certainly senescence is a medical problem. From general population genetics I knew something about inherited disorders. These are quite different kinds of medical problems, but all of them are susceptible to evolutionary interpretations, in ways that it seems to me would benefit the practice of medicine. The more I got to thinking about it, and talking to Randy Nesse about it, the more I realized that there is no kind of medical problem for which the theory of natural selection will not be relevant, for curing or preventing a disease.

One of Paul's most important insights is that AIDS is probably not a new disease, in the sense of HIV being a new pathogen. What we're dealing with is a pathogen that has rapidly evolved a much higher level of virulence because of its environmental circumstances. It may have been an organism that, prior to two or three decades ago, was transmitted primarily from parent to offspring—and maybe rarely between sex partners—and therefore the evolutionary factors acting on its virulence necessarily kept it very nonvirulent. Individuals with this virus had to survive long enough to reproduce, or the virus wouldn't be transmitted.

Now, take people with this virus and move them into a completely different social situation, in which families are disrupted and men are being served mainly by prostitutes who are dealing with hundreds of men per year. You now have a situation in which the opportunity for the transmission of the disease to another individual no longer depends upon the long-term survival of the individual that has it. Therefore the restraints on its virulence are removed. Within an individual, the more virulent the strain, the better it will do, because the more virulent the strain the more of that particular virus there will be for transmission to the next individual. We've shifted the balance of selection on this virus from mainly between individuals—between hosts—to within hosts. Within hosts, there's normally selection for increased virulence. Suddenly the virulence of the HIV went way up. This is just one example. There are many, many examples of human activities that influence the evolution of virulence in our pathogens.

There are many other ways in which evolutionary ideas can be brought to bear on medicine—for instance, in dealing with the mismatch between our evolved adaptations and the environment in which we now find ourselves. This mismatch is probably the main source of medical problems today.

In twenty or thirty years, medical students will be learning about natural selection, about things like balance between unfavorable mutations and selection. They will be learning about the evolution of virulence, of resistance to antibiotics by microorganisms, they will be learning about human archaeology, about Stone Age life, and the conditions in the Stone Age that essentially put the finishing touches on human nature as we now have it. These same ideas then will be informing the work of practitioners of medicine, and the interactions between doctor and patient. They'll be guiding the medical research establishment in a fundamental way, which isn't true today. At the rate things are going, this is inevitable. These ideas ought to reach the people who are in charge—the doctors and the medical researchers—but it's even more important that they reach college students, especially future medical students, and patients who go to the doctor. They'll have questions to ask that doctor, who will have to have answers. I hope this set of ideas produces a certain amount of bottom-up influence on the medical community, via students and patients. But I hope also that there's some top-down influence—that it will be influencing the faculties in medical schools and the researchers on human disease.

STEPHEN JAY GOULD: George Williams is a very important man. He's a quiet, gentle man who has had enormous influence on evolutionary theory since the 1960s, particularly through *Adaptation and Natural Selection*, in 1966, which was largely a critique of the false logic in forms of group selectionism then current and a defense of a fairly hard-edged strict Darwinian view based on individual selection. It was a methodological argument; he didn't say that group selection is impossible in principle, he just said that the arguments heretofore adduced were fallacious, and in that context one must begin (and here I don't agree with him philosophically) at the reduced, or lowest, level—namely, Darwinian competition among organisms—and not claim that selection is operating among any higher-level entities, like groups or species, unless you have to. If everything can be explained by organisms, let it go by organisms. Very influential book. He's always been at the forefront of theoretical clarity in the field.

RICHARD DAWKINS: I have enormous regard for George Williams; I see him as an immensely wise figure in my field. And he has been—belatedly—enormously influential. The essence of *The Selfish Gene*, which came out in 1976, is contained in a couple of paragraphs in Williams' *Adaptation and Natural Selection*. I had not read it when I independently realized the same thing. His book has been a colossal influence for the good in the development of evolutionary theory and is now widely looked up to as such; it wasn't looked up to so much to begin with, but it's one of those (I think you call them) slow burners, whose influence develops rather late. I have huge regard for him.

LYNN MARGULIS: The only book of his I have read is *Adaptation and Natural Selection*. He makes a contribution in enlightening those who don't understand the basic idea of evolution. Most people don't understand the consequences of a simple fact: reproduction in mammals is obligatorily sexual—although, for life generally, no intrinsic requirement for reproduction to be correlated with sex exists. But human behavior with its sexual reproductive imperatives can be understood as a function of the evolutionary past. People didn't connect evolutionary thinking and mammalian behavior. Williams is credited for recognizing the importance of reproduction in mammalian behavior. He's communicating a scientific truth in a resistent cultural milieu. The fact that he has few articulate predecessors enhances the importance of his work.

STEVEN PINKER: In my mind, George Williams is one of the most brilliant writers in the history of science. His 1966 book *Adaptation and Natural Selection* was way ahead of its time. In the first part of his argument, Williams was castigating contemporary biologists, pointing out that many of their explanations were shoddy because they were invoking natural selection as an explanation for every beneficial trait in sight. No matter what they looked at in an organism, they could come up with a story as to why it was to the benefit of the organism, the species, the ecosystem, the community, or the planet. Williams carefully dissected the concept of natural selection, delineating where it should and should not be applied. He noted that not everything that's adaptive is an adaptation in the technical sense. If a fox's feet tamp down a path in the snow, and that helps the fox get to the henhouse, it doesn't mean that the feet of the fox are an adaptation to tamping down snow.

The second part of Williams' argument is that even though natural selection can't explain every trait, there are some traits for which it's the only scientific explanation. These are the traits that show signs of complex adaptive design. The hand, the eye, the heart, the skin—all are extremely improbable arrangements of matter. General laws of growth or the accidents of genetic drift couldn't possibly explain the precise arrangements of muscles and bones and tendons that give us a usable hand—or, for that matter, explain why something like the apple has seeds inside it as opposed to something else. For any biological structure that looks as if it's engineered for a purpose, natural selection is the only known scientific explanation, because it's the only physical process that can result in complex systems that achieve some improbable goal.

Williams presented both halves of the argument. Some traits you shouldn't use natural selection for; some traits you have to use natural selection for. A lot of the work of biology, the day-to-day work, is examining complex features of organisms and trying to figure out whether they could have arisen as a by-product of something else or show clear-cut signs of having been designed for some purpose. Dawkins' book *The Blind Watchmaker* is in large part a lucid extension and popularization of both halves of Williams' original idea. Much in the writings of Stephen Jay Gould and his colleague Richard Lewontin emphasizes the first half and ignore the second half.

NILES ELDREDGE: I remember the English evolutionary geneticist John Maynard Smith remarking to me that he was astonished to find out that George Williams wasn't in our National Academy. Williams finally got elected in 1993. When I visited him in Stony Brook in the mid-1980s, he told me he was having a hard time getting grant support for his research, and I couldn't believe that. The two thoughts converged, because George really is the most important thinker in evolutionary biology in the United States since the 1959 Darwin centennial. It's astonishing that he hasn't gotten more credit and acclaim. He's a shy guy, but a very nice guy, and a very deep and a very careful thinker. I admire him tremendously, even though we've been arguing back and forth for years now.

His best book was the 1966 *Adaptation and Natural Selection*. I have more problems with *Sex and Evolution*. He has misunderstood some of the things we're trying to say, in a way that sometimes I find frustrating, shall we say. I don't think it's so much that he's being perverse as that I'm having a hard time getting through to George on certain things right now. But nonetheless the respect that we all have in the field for George is there. The guy is solid gold.

DANIEL C. DENNETT: As other people have said, George Williams is the Abraham Lincoln of his field. He has a wonderful, laconic, pithy way of talking, and he seems to be an amazingly astute and clearheaded thinker. Reading George Williams showed me for the first time how hard it is to be a good evolutionary thinker, and how easy it is to make simple mistakes. Again and again, Williams issues his pithy little correctives to otherwise superficially good ideas and just calmly, firmly, wipes them out. Then you realize that this is a harder game to play than any of us realize, and George plays it better than anybody else in the world.

His main contribution, of course, was blowing the whistle loud and clear on the idea of "good for the species." In his 1966 book, he saw that Wynne-Edwards'—and others'—ideas, which were very familiar fare in the textbooks and popular treatments of evolution, had to be wrong. This was a wake-up call. Williams pointed out that it's not "What's good for the species is good for the organism (or vice versa)"; it's "What's good for the gene is good for the gene." Usually, other things being equal, what's good for the gene is good for the organism—and thus, you might say, for the species. But the gene is in the driver's seat.

Chapter 2

STEPHEN JAY GOULD
"The Pattern of Life's History"

STUART KAUFFMAN: *Steve is extremely bright, inventive. He thoroughly understands paleontology; he thoroughly understands evolutionary biology. He has performed an enormous service in getting people to think about punctuated equilibrium, because you see the process of stasis/sudden change, which is a puzzle. It's the cessation of change for long periods of time. Since you always have mutations, why don't things continue changing?*
You either have to say that the particular form is highly adapted, optimal, and exists in a stable environment, or you have to be very puzzled. Steve has been enormously important in that sense.

• • •

STEPHEN JAY GOULD *is an evolutionary biologist, a paleontologist, and a snail geneticist; professor of zoology at Harvard University; MacArthur Fellow; author of, among others,* Ontogeny and Phylogeny *(1977),* The Mismeasure of Man *(1981),* The Flamingo's Smile *(1985),* Wonderful Life *(1989), and* Bully for Brontosaurus *(1992).*

STEPHEN JAY GOULD: There is no progress in evolution. The fact of evolutionary change through time doesn't represent progress as we know it. Progress is not inevitable. Much of evolution is downward in terms of morphological complexity, rather than upward. We're not marching toward some greater thing.

The actual history of life is awfully damn curious in the light of our usual expectation that there's some predictable drive toward a generally increasing complexity in time. If that's so, life certainly took its time about it: five-sixths of the history of life is the story of single-celled creatures only.

I would like to propose that the modal complexity of life has never changed and it never will, that right from the beginning of life's history it has been what it is; and that our view of complexity is shaped by our warped decision to focus on only one small aspect of life's history; and that the small bit of the history of life that we can legitimately see as involved in progress arises for an odd structural reason and has nothing to do with any predictable drive toward it.

I'm working on an incubus of a project on the structure of evolutionary theory, an attempt to show what has to be altered and expanded from the strict Darwinian model to make a more adequate evolutionary theory.

Basically, there are three themes. The first is the hierarchical theory of natural selection—selection operating on so many levels, both above and below. Richard Dawkins, who still wishes to explain virtually everything at the level of genic selection, is right about one thing; gene selection does operate. He's wrong in saying that it's *the* source of evolution; it's *a* source. I don't know what the relative strength and power of the levels are—it depends on the particular problem—but gene selection is not the dominant one, by any means. It certainly happens; it may be responsible for the increase in the number of multiple copies of some kinds of DNA within evolutionary lineages, for example; it's responsible for some things.

The second theme is the extent to which strict adaptationism has to be compromised by considering the developmental and genetic restraints at work upon organisms, when you start considering the organism as a figure that pushes back against the force of natural selection. The best way to explain it is metaphorically. Under really strict Darwinism (Darwin is not a strict Darwinian), a population is like a billiard ball: you get a lot of variability, but the variability is random, in all directions. Natural selection is like a pool cue. Natural selection hits the ball, and the ball goes wherever selection pushes it. It's an externalist, functionalist, adaptationist theory. In the nineteenth century, Francis Galton, Darwin's cousin, developed an interesting metaphor: he said an organism is a polyhedron; it rests on one of the facets, one of the surfaces of a polyhedron. You may still need the pool cue of natural selection to hit it—it doesn't move unless there is a pushing force—but it's a polyhedron, meaning that an internal constitution shapes its form and the pathways of change are limited. There are certain pathways that are more probable, and there are certain ones that aren't accessible, even though they might be adaptively advantageous. It really behooves us to study the influence of these structural constraints upon Darwinian and functional adaptation; these are very different views.

The third theme is the extent to which a crucial argument in Darwinism—namely, that you can look at what's happening to pigeons on a generational scale and extrapolate that into the immensity of geological time—really doesn't work, that when you enter geological time there are a whole set of other processes and principles, like what happens in mass extinctions, that make the extrapolationist model not universal.

I'm attempting to marry those three themes—hierarchical selection, internal constraint, and the immensity of geological time—into a more adequate general view of evolutionary theory.

I should say that geological time is in there because it's so essential to strict Darwinian theory that you be able to use the strategy of bio-uniformitarian extrapolation; in other words, that you be able to see what happens in local populations, and then render the much larger-scale events that occur through millions of years to much larger effect by accumulation of these small changes through time. If, in the introduction of the perspective of millions of years, new causes enter that couldn't ever be understood by studying what happens to pigeons and populations for the moment, then you couldn't use the Darwinian research strategy. That's why Darwin himself was so afraid of mass extinction and tried to deny the phenomenon. The geological stage is really a critique of the uniformitarian, or extrapolationist, aspect of Darwinian thinking.

Richard Lewontin is my population-genetics colleague at Harvard, probably the most brilliant man I've ever had the pleasure of working with. We teach a "Basics of Evolution" course together. In 1978, there was a symposium on adaptation held by the Royal Society of London. It was a very pro-adaptationist symposium; that's the British hang-up, after all. I think John Maynard Smith was one of the organizers. Dick was invited to present a contrary view, because—particularly after the publication in 1975 of E.O. Wilson's *Sociobiology*, which is so strongly adaptationist—Dick had been quite vocal in his doubts about the adaptational parts.

Clearly, there's a lot of adaptation in nature. Nobody denies that the hand works really well, and the foot works well, and I don't know any way to build well-adapted structures except by natural selection. I don't have any quarrel with that, and I don't think any serious biologists do. But adaptationism is the hard-line view—which has been so characteristic of English natural history since Darwin—that effectively every structure in nature (there are exceptions of course) needs to be explained as the result of the operation of natural selection; that if we're not absolutely optimal bodies—because clearly we're not—we're at least maximized by natural selection.

Darwinian biologists will use it as the strategy of first choice. If you see a structure in a flower or in a mole, and you don't know what it's for, the first thing you assume is that it was built by natural selection for something, and your job is to figure out why it's there—the "why" being "What is it good for?"—because once you know what it's good for, then you know why natural selection made it. Although this is a technique that often works, it's inadequate in so many cases that it just doesn't suffice as a general strategy, the main problem being that many structures are built for other reasons that have nothing to do with natural selection. For example, they can arise as side consequences of other features that might have adaptive benefit. Having been built for other purposes, they may then prove useful; they can be coopted secondarily for utility. The bird's wing did not evolve for flight. If you want to know why it's there, seeing a bird fly isn't going to help you, because 5 percent of a wing doesn't fly. It must have been built originally for some other function.

Take the human brain. Most of what the human brain does is useful in a sense—that is, we make do with it—but the brain is also an enormously complex computer, and most of its modes of working don't have to be direct results of natural selection for its specific attainments. Natural selection didn't build our brains to write or to read, that's for sure, because we didn't do those things for so long.

Anyway, the Royal Society asked Dick to write a piece for the 1978 symposium. I had developed my own doubts about adaptationism, for a host of reasons. Part of it came from working on random models of phylogeny with Dave Raup and Tom Schopf and Dan Simberloff in the early seventies and coming to realize how much of an apparent pattern could be produced within random systems. Part of it came from writing my first book, *Ontogeny and Phylogeny*, in 1977, and coming in contact with the great German and French continental literature on structural, or nonadaptational, biology. That's the continental tradition as much as adaptationism is the English tradition. I also had been unhappy with the overuse of adaptation in sociobiological literature, so I had a whole variety of reasons to agree with Dick on those subjects.

Dick was going to be the one nonadaptationist speaker at the symposium. In fairness, he was going to give the last speech, and it was certainly given prominent coverage; the English are nothing if not fair.

Dick doesn't like to fly, and he had no particular desire to go there, and since we had pretty consonant views and I wanted to go to England anyway, we decided to write a joint paper. In fact I wrote virtually all of it. He was very busy, and I would be giving the paper anyway. The paper is a general critique of full-scale adaptationism, or panadaptationism. It's not an attempt to trash Darwinian natural selection, which obviously happens; it's an attempt to argue that adaptationism, or the notion that Darwinian selection is effectively responsible for everything in the form of organisms, just will not work.

One of the main reasons I'm proud of that paper is that I do believe in interdisciplinary perspectives and—as an essayist, particularly—the use of examples from other fields. The paper succeeded because I used a fairly arresting strategy of argument, by beginning with an architectural example.

The paper is called "The Spandrels of San Marco and the Panglossian Paradigm; A Critique of the Adaptationist Program." I began by talking about the spandrels under the domes of the cathedral of San Marco. I had been in Venice a few months before, and I had stood under the dome in San Marco, and I had worked out this argument for myself, and it was very enlightening to me. It helped me to see what's wrong with the adaptationist paradigm.

Here's the situation: You decide to build a church by mounting a circular dome on four rounded arches that meet at right angles. I'll accept that as an analog of adaptation; that's an engineering design that works. But once you do that, you have four tapering triangular spaces where any two arches meet at right angles. The spaces are called spandrels—or pendentives, but the more general architectural term is spandrels. They're spaces left over.

No one can claim that the spandrels under the dome are adaptations for anything. I suppose it's a good idea to put some plaster there—otherwise the rainwater is going to come in—but the fact that they're tapering triangular spaces is a side consequence of the adaptive decision to mount the dome on four arches. It's space left over. It's a side consequence; it isn't an adaptation in itself.

When I looked at these spandrels, I realized that every set of spandrels—there are six in San Marco—had a very sensible iconography linked with the dome. Under the main dome, for example, there are four evangelists in the spandrels. Four spandrels, four evangelists. Under each of the four evangelists is one of the four biblical rivers—the Tigris, the Euphrates, the Nile, and the Indus—and they are personified as a man, and the man holds an amphora, a water jar, and he pours water onto a single flower in the tapering triangular space below. It's a beautiful design. But no one would argue that the spandrels exist to house the evangelists. The spandrels are nonadaptive, side consequences. Since they are there anyway, you might as well fill them with useful and sensible structures.

Many biologists would say, "Well, of course, that's right. We know there are spandrels, or bits and pieces, left over, but they're just nooks and crannies, funny little corners; they don't have any importance." But that's not true; the fact that something is secondary in its origin doesn't mean it's unimportant in its consequences. Those are entirely separate subjects.

Spandrels often turn out to be more important, in terms of the consequences in history of a structure, than the actual immediate reasons for their having been there in the first place. For example, the dome of San Marco is radially symmetrical; there is no reason to ornament the dome in four-part symmetry for structural reasons, yet every dome but one in San Marco is ornamentally structured in four-part symmetry, in harmony with the spandrels below. The spandrels are not just nooks and crannies; they actually determine the iconographic program of the dome itself. Just as with the human brain: most of what the brain does are probably spandrels—that is, the brain got big by natural selection for a small set of reasons having to do with what is good about brains on the African savannas. But by virtue of that computational power, the brain can do thousands of things that have nothing to do with why natural selection made it big in the first place, and those are its spandrels.

Because I began this paper with an architectural example, no one would confute it, because it wasn't a threat to their conventional thinking. If I'd started with an organic example, it would have raised the hackles of all the people trying to be strict Darwinians.

Arthur Cain was the summer-up of the whole session. There's a line in Durrell's *Alexandria Quartet* where the narrator, Pursewarden, says that he's a Protestant in the only meaningful sense of the term—that he likes to protest. Well, moderators are supposed to be moderate, I suppose. Arthur Cain was not. He devoted his entire summary of the conference to a vitriolic attack on this paper, essentially saying that Dick and I knew that adaptation was true because we had to, because it obviously is true. Arthur said that we had attacked it because, although we knew it to be true, we so disliked the political implications of sociobiology, which is based on it, that we abrogated our credentials as scientists.

That was so off the wall that it was just amazing. When I got up to give my re-reply, the second coordinator of the conference was standing in front of the podium—which had the motto of the Royal Society, "Nullius in verba," on it—and I asked him to step aside. He was annoyed: why was I asking him to move, that was not fair. But he later realized why I had. Now, I'm stupid about certain things that scientists are supposed to be good at. I'm not particularly quantitative; I'm numerate but not innovative. I'm not a great experimentalist. But I pride myself on having immersed myself in Western culture and having learned some languages, and knowing certain aspects of humanism that many scientists don't take up.

I asked him to step aside, and I said that I thought Arthur had been entirely wrong, that he'd completely misunderstood the motives of my talk, and that I was doing nothing but trying to uphold the motto of the Royal Society, which had sponsored this meeting. The reason that was an effective strategy was that I knew that most people, most members, didn't know what the motto "Nullius in verba" meant. It looks like it means "Words do not matter" or "Do not pay any attention to words," since *nullius* means "nothing" and *verba* is "word." So most people think it means that words mean nothing and you have to do the experiment.

But *nullius* is genitive singular; it can't mean that. It means "of nothing" or "of no one." I knew what the motto meant. I knew that it was a fragment of a statement from Horace—a famous quotation from a poem, in which he says, "I am not bound to swear allegiance to the dogmas of any master." *Nullius addictus jurare in verba magister.* It's "Nullius in verba," or "In the words of no (master)." It's just a fragment from a larger line.

"That's all I'm doing," I said. "I'm saying that we are not bound to swear allegiance to the dogmas of any master; I'm here to present an alternative viewpoint that's consistent with your own society. How can you castigate me?"

The paper gets a lot of citations, but I don't know how many of its citations mean that it was actually used. In the game of citation analysis, you know that there are a certain number of citations that are, in a sense, honorary; that is, people will write a paper in which they want to support an adaptationist's perspective, and they feel that in fairness they have to cite at least one thing to show they know there's an opposing literature. The spandrels paper is the classic one, so they cite it. Whether or not they actually take it seriously I don't know. But it's become the standard source of a broader view of the causes of evolutionary form.

The paper provides a context for my current views on constraint—the importance of geometric and historical constraint, as opposed to a strictly adaptationist view of the world. The "exaptation" argument arises very much out of the spandrels principle, and I wish I'd developed the word when I wrote the paper. There's a problem—most Darwinians don't acknowledge it, since it doesn't work out as a problem for them—because "adaptation," as the word is used, has two distinctly different meanings. It's the process whereby a structure is designed by natural selection for a use, but often the word is also used for the structure itself. I have my foot here. It works well. Is it an adaptation, simply because it works well? Strict Darwinians don't have a problem using the same word both for the structure that works well and for the process that gets you there, because they think that the process is the only way you can get the working structure.

Under the spandrel principle, you can have a structure that is fit, that works well, that is apt, but was not built by natural selection for its current utility. It may not have been built by natural selection at all. The spandrels are architectural by-products. They were not built by natural selection, but they are used in a wonderful way—to house the evangelists. But you can't say they were adapted to house evangelists; they weren't. That's why Elisabeth Vrba and I developed the term "exaptation." Elisabeth is a paleontologist at Yale University, who has collaborated with both Niles Eldredge and me, and who did the most interesting work on punctuated equilibrium.

Exaptations are useful structures by virtue of having been coopted—that's the "ex-apt"—they're apt because of what they are for other reasons. They were not built by natural selection for their current role. Strict Darwinians cannot deny the principle. Their usual response is to say that it's minor, just a gloss, exaptations are rare, they're just nooks and crannies, they're not important. But in the spandrels argument it's essential that they are important. Just because something arises as a side consequence doesn't condemn it to secondary status.

Arthur Cain brought up the subject of political implications. In a sense, I brought it on myself, but I'll defend how it happened. Niles

Eldredge and I wrote the first punctuated-equilibrium paper in 1972. I wrote a follow-up in 1977, in which I tried to analyze some of the theory's social and psychological sources, because they're in every theory of gradualism, and I had tried to argue that gradualism is a quintessential notion of Victorian liberalism. I thought it would be so ridiculous and—to use a biblical term—vainglorious to claim that gradualism, at least in part, was not a truth of nature but recorded a social context, and then to argue that "punctuated equilibrium is true; it's just a fact of nature." There obviously had to be a social context for punctuated equilibrium, too. I thought it only fair to write about what might have been some of the sources of punctuated equilibrium, and since there's a long tradition in Hegelian and Marxist thought for punctuational theories of change, it was clearly not irrelevant that I had been brought up by a Marxist father. I'd learned about these things.

That's not the reason the punctuated-equilibrium theory exists— if only because Niles developed most of the ideas, and he didn't have any such background. But it is relevant that I, rather than someone else, thought of it, in that my own background is probably a relevant fact. It was necessary for me to say that; it would have been absurd to claim that gradualism is politically influenced but punctuated equilibrium is a fact of nature. People seize upon that one statement.

Historians of science make a distinction between what they call context of justification and context of discovery, and it's fair enough. There's a logic of justification, which is independent of the political and social views of the people who develop the ideas. But if you want to ask why certain people develop ideas rather than other people, and why they develop them in this decade rather than that decade, then for those questions, which are about context of discovery rather than context of justification, surely the personal side is very relevant; it has to be explored and understood. But it has very little bearing on whether the idea is right or not. The fact that I learned Marxism from my father may have predisposed me toward being friendly to the kind of ideas that culminated in punctuated equilibrium; it has absolutely nothing to do with whether punctuated equilibrium is true or not, which is an independent question that has to be validated in nature.

Within a profession, certain issues can become very big which, if seen from the outside, might not seem so. For instance, in evolutionary theory, on the outside the only issue might be whether evolution is true or not. That's the big one! On the inside, of course, everyone knows that evolution is true; the issue is how it occurs. The main difference between Richard Dawkins and myself has to do with the agency of natural selection, and its power, and the degrees of adaptation that it produces. Within the field, these questions define the essence of Darwinism; outside the field, they might seem smallish. It is just a question of perception.

Richard wants natural selection to be effectively all-powerful, at least when you are dealing with the phenotypes—the forms of organisms. He wants the locus of that selection to be genes. I maintain that natural selection works on a hierarchy of levels simultaneously, of which genes are one and organisms are another, and that you also have higher units, such as populations and species, at which selection is very effective, and the end result is not always, by any means, adaptation—particularly when you see the process unfolding in millions of years of geological time.

No matter how effective adaptive change might be in the moment, when you start translating that and any other process into millions of years, it doesn't work out that the history of life is under adaptative control, because you have to get through these largely random and highly contingent mass-extinction events, as well as new species arising by punctuated equilibrium. Long-term success in clades is the function of speciation rate, which has very little to do with the morphologies that are built by natural selection. So Richard's and my whole views of evolutionary mechanics are very different, but to the outsider, who may only be concerned with whether evolution happens or not, we probably seem to be pretty similar, because we are both evolutionists.

I would call Richard's approach hyper-Darwinism. The brilliance of Darwin's argument, and the radical nature of it, lies in changing the focus of explanation. Before Darwin, people thought that organisms were well-designed because the highest-order force was doing it directly. There was a benevolent, creative God who made it that way. The brilliance of Darwin is that he beat the level of explanation down to organisms, saying that organisms are well-designed as a side consequence of their struggle for individual reproductive success. It is a deliciously radical argument. Instead of an all-wise, benevolent, purposeful God, what you have are organisms struggling for personal advantage—which seems to be the moral opposite, except that there is no morality in nature—and as a side consequence you get good design of organisms.

Richard has taken that posture of trying to beat the level of explanation down, and has carried it to its ultimate extreme: it's not even the organisms that are struggling, it's only the genes. The organisms are "vehicles." That's his pejorative word; most of the profession calls them "interactors," which is less pejorative. The only active agents in Richard's worldview are genes. He's wrong. If you read the British philosopher Helena Cronin's book *The Ant and the Peacock*, she argues that the whole profession has been transformed by this idea. Whatever my personal point of view might be, her claim is sociologically wrong in a purely factual or Gallup Poll sense. Not many people take this view seriously. A lot of people like it as a metaphor for explanation. But I think that very few people in the profession take it seriously, because it's logically and empirically wrong, as many people, both philosophers and biologists have shown—from Elliott Sober to Richard Lewontin to Peter Godfrey Smith.

Richard is basically wrong, because organisms are doing the struggling out there. If organisms could be described as the additive accumulation of what their genes do, then you could say that organisms are representing the genes, but they're not. Organisms have hosts of emergent characteristics. In other words, genes interact in a nonlinear way. It is the interaction that defines the organism, and if those interactions, in a technical sense, are nonadditive—that is, if you can't just say that it's this percent of this gene plus that percent of that gene—then you cannot reduce the interaction to the gene. This is a technical philosophical point. As soon as you have emergent characteristics due to nonadditive interaction among lower-level entities, then you can't reduce to the lower-level entities, because the nonadditive features have emerged. These features don't exist until you get into the higher level. His argument is wrong. It's not just a question of being inadequate. It's wrong.

Admittedly—again, in a sociological sense—it's enormously appealing. When you realize what Darwin did, which was to break down the explanation from the benevolent God to the struggling organism, the notion that you might break the explanation down further, to the struggling gene, has a certain reductionist appeal. But if you surveyed the profession, although not all of them would necessarily agree with me about hierarchical selection, most would say that Darwin was right and selection is primarily on organisms, which has always been the traditional view.

Gene selectionism was never a paradigm that attracted large numbers. What did happen was that the generation before Dawkins, culminating in 1959, had a form of very strict Darwinian adaptationism, a more classic, organism-centered Darwinian approach that wasn't by any means totally wrong but was much too restrictive. It did become a ruling view within evolutionary theory, and to some extent we're still fighting it, in talking about large-scale, macroevolutionary changes as not being fully extrapolatable out of the adaptive struggles of organisms and populations.

I might be on the periphery of orthodoxy, but I certainly think natural selection is an enormously powerful force. Darwin's canonical form of it—that is, selection operating on individual bodies via the struggle for reproductive success—just isn't capable, by extrapolation, of explaining all major patterning forces in the history of life. Whereas it's vital for strict Darwinism that you do accept such a view. You'll always have a little bit here and there for other things, to be sure, but unless you can argue that Darwinian selection on bodies is, by extrapolation, the cause of evolutionary trends and of the major patterns of waxing and waning of groups through time, then you don't have a fully Darwinian explanation for life's history.

I see Dawkins in a dual sense. On the one hand, he's the best living explainer of the essence of what Darwinism is all about. That part's very good. He's a kind of old-fashioned, nineteenth-century, almost atheistic scientific rationalist. The other side is the strict Darwinian zealot, who's convinced that everything out there is adaptive and is all a function of genes struggling. That's just plain wrong, for a whole variety of complex reasons. There's gene-level selection, but there's also organism-level and species-level. Those are his two sides: the professional true believer, on the one hand, and the excellent explainer of a worldview, on the other.

I'd question Richard on the issue of gene-level selection and why he thinks that the issue of organized adaptive complexity is the only thing that matters. I'm actually fairly Darwinian when it comes to the issue of so-called organized adaptive complexity, but there's so much more to the world out there. Why does he think that adaptation in that sense is responsible for interpreting everything in the history of life? Why does he insist on trying to render large-scale paleontological patterns as though they were just grandiose Darwinian competitions? They aren't. He has this blinkered view in which the classic Darwinian question of adaptation is somehow becoming coextensive with all of evolutionary theory.

Richard and I are the two people who write about evolution best. He writes about microevolutionary theory, in a way I disagree with. I focus on the pattern of life's history and its relationship to evolutionary theory. I treat the fossil record and write about macroevolutionary theory, which he doesn't like. He writes on the nature of adaptation and on evolutionary theory in its traditional small-scale immediacy, and I write about the large-scale history of life.

Whether or not Darwin would be a Darwinist today, in the way the word is used, is so hard to say, because you have to make inferences about his mental flexibility. Given the set of ideas that he himself promulgated, I think he would, because his tendency in argument was always to try and stretch natural selection on bodies to cover cases. He was willing to allow a few very circumscribed exceptions, like his invocation of group selection for the evolution of human moral behavior—an important exception, to be sure, because we care about human moral behavior. But he circumscribed it in such a way that it could apply to no other species, because he invoked a group-selection mechanism that could work only in highly cognitive species that are sensitive to the "praise and blame of their fellows"—those are his words—and we're the only such species. So therefore he set up the exception in such a way as to marginalize it; it's an important one, because it's about us and we care about us, but it's not important in the full realm of nature.

On the other hand, if you want to speculate psychologically, Darwin was an enormously flexible, brilliant, and radical thinker, so I suspect that when he learned about asteroidal impact and mass extinction and maybe even punctuated equilibrium, he would be open. I doubt that he expected that a hundred years after his death things would be exactly as he had left them.

STUART KAUFFMAN: Steve is extremely bright, inventive. He thoroughly understands paleontology; he thoroughly understands evolutionary biology. He has performed an enormous service in getting people to think about punctuated equilibrium, because you see the process of stasis/sudden change, which is a puzzle. It's the cessation of change for long periods of time. Since you always have mutations, why don't things continue changing? You either have to say that the particular form is highly adapted, optimal, and exists in a stable environment, or you have to be very puzzled. Steve has been enormously important in that sense.

Talking with Steve, or listening to him give a talk, is a bit like playing tennis with someone who's better than you are. It makes you play a better game than you can play. For years, Steve has wanted to find, in effect, what accounts for the order in biology, without having to appeal to selection to explain everything—that is, to the evolutionary "just-so stories." You can come up with some cockamamie account about why anything you look at was formed in evolution because it was useful for something. There is no way of checking such things. We're natural allies, because I'm trying to find sources of that natural order without appealing to selection, and yet we all know that selection is important.

MARVIN MINSKY: What I love about Stephen Gould is his ability both to research and to explain the possible evolutionary pathways that might have led to what we see in particular cases. His explanations and hypotheses are constructed from the most diverse kinds of evidence, by combining both general principles and particular details from many different fields. It's a wonder to see so many aspects synthesized at all—and perhaps more of a wonder to see them described with such beauty and clarity.

NILES ELDREDGE: Steve and I are like brothers, and when we get together we mostly like to talk about the things we disagree on, but of course the rest of the world is hard pressed to see how we differ on anything at all. Yet we do. That, to us, is the most interesting stuff. When we first wrote the punctuated-equilibrium papers, I thought it was more about mode and Steve thought it was more about tempo, using the two phrases from George G. Simpson's *Tempo and Mode in Evolution*. We had a different take on what it all meant. I think to some degree we probably still do.

Steve is prodigious. I never met somebody who was so smart who worked so hard. He is a marvelous scholar. I have never found anybody who could grasp the essence of an issue so quickly, either. He was an inspiration to me when we were graduate students, because he showed that it was possible—and, indeed, it was almost an obligation—for young people to think critically, to think theoretically, and to publish. He showed the way.

The downside of being associated with Steve, of course, is that sometimes you feel like you are standing in a shadow, that you're one of the also-rans. But I've benefited far more than I've suffered from being associated with Steve, and I think we're closer now than perhaps ever before.

MURRAY GELL-MANN: Stephen Jay Gould and I collaborated in consulting on, and obtaining signatures for, an amicus-curiae brief for the Supreme Court in *Edwards v. Aguillard*, which was the Louisiana creationism case. We called on the Supreme Court to declare that it was unconstitutional to force science teachers in Louisiana to devote equal time to the doctrine of creationism if and when they taught about evolution, since evolution is the scientific account of how life developed on earth and creationism is an idea that no one would believe today who is not starting from some form of fundamentalist religious dogmatism. Our side won, seven to two.

FRANCISCO VARELA: I feel very close to many of the fundamental ideas that Steve Gould has come up with, and I've learned from his critique of the adaptationist program, in the famous paper he wrote with Lewontin.

I've been fighting for many years, in the case of the operation of the brain, to make the point that the brain is not an information machine that picks up information and creates an optimal representation of what's out there. The whole story is quite otherwise. There is an absolutely identical analogy with evolution. In the traditional, simplistic Darwinian view, adaptation is some form of optimal fit with a given world. What Gould is saying is that the adaptationist idea that there's an ideal world to which species fit is just nonsense; that there is instead an intrinsic story, an internal story, to evolution—or intrinsic factors, as they are called now—which shapes the niche, and the form of the species, just as much. This is the same thing I'm saying about the brain—or about the immune system, for that matter. His critique of the post-Darwinian adaptationist view is very much in resonance with my own work.

That's saying nothing about something else I admire enormously: Gould's ability to communicate ideas to the large public. That's his unique genius. Anybody who has read, for example, *Wonderful Life*, realizes that he can take something which is obscure and abstruse, and not only make it relevant to the large public, but actually in the same stroke produce a new reading of a fundamental chapter of biology.

With regard to the Dawkins-Gould debate, if I wanted to be brutal I would say that Gould is right and Dawkins is wrong.

J. DOYNE FARMER: Stephen Jay Gould is an excellent writer and a clear thinker, and he has a real gift for writing about scientific issues and providing enough personality and drama so that nonscientists can get excited about what he's saying: he's perhaps the Herbert Spencer of our day. He doesn't know complexity theory and he doesn't care. My guess is that he wouldn't see much value in something like artificial life.

Gould is from the old school. He's a biologist, he's not educated mathematically. He may have a perfectly clear concept of what physics is, but he certainly isn't in any sense attempting to achieve the levels of abstraction or generality for evolution or evolutionary biology that have been achieved in physics.

STEVEN PINKER: In Ernst Mayr's authoritative history of biological thought, he notes the irony that paleontologists were the biologists most skeptical of natural selection. Presumably it's because paleontologists study organisms after they've turned into rocks, and their first concern can't be how stomachs work, or how eyes work, or how the visual circuitry of the brain works. The evolutionary geneticist John Maynard Smith has suggested that Gould fits into this tradition in much of his writing, because natural selection doesn't answer the first questions that paleontologists face—namely, what are the grand patterns in the history of life: why does one kind of animal replace another over a span of tens of millions of years?

To be fair, there used to be a widespread idea that natural selection could explain just such facts. The mammals succeeded the reptiles because in some way they were better adapted, or fitter. Gould has eloquently shown some of the problems of this application. But it's something that modern Darwinians, like Maynard Smith, Richard

Dawkins, and George Williams, wouldn't claim to begin with. They'd be happy to concede that many macroevolutionary phenomena can't be explained by natural selection—a clear example being the possible extinction of the dinosaurs because of a collision between the earth and an asteroid or a comet. But biologists outside of paleontology study the complex functioning of individual organisms, and that's why they're much more likely to appreciate the power of natural selection.

Many scientific debates are like the blind men and the elephant: different people are interested in different aspects of the problem. Scientists will imagine that they're in sharp disagreement with other scientists, when they're merely studying something else. Gould's criticisms of Dawkins, Helena Cronin, and those he calls sociobiologists are a bit like that: those people are using natural selection to answer questions about complex form and behavior, where natural selection is required, and he points to areas of biology like mass extinctions or differences in banding patterns on snails, where it's not required. In fact, Dawkins would be the first to agree that there are certain things for which natural selection is not the best explanation. What Dawkins says—quite convincingly, in my mind—is that the kinds of questions that a physiologist or an anatomist or an ethologist or a cognitive scientist is interested in are the kinds of questions that you do need natural selection for.

I greatly admire Steve Gould's writings, and I've learned an enormous amount of biology from them. And I agree with some of his leitmotifs, such as the lack of progress in evolution, the importance of understanding phylogeny as a tree rather than as a ladder, and the importance of contingent historical events in evolution. But there are others that I have problems with. For one thing, I don't think he fully acknowledges the complexity of everyday unconscious mental processes. He has drawn misleading analogies about how the mind might be like a computer or a general-purpose learning device. He suggests that just as a computer can play tic-tac-toe as well as calculate a company's payroll, the brain could have been designed for one thing and used for other things. But that's not quite right. You can't take a computer out of the box and have it both compute a company's payroll and play tic-tac-toe. Someone has to have programmed it specifically for both tasks, so the analogy falls apart, even in the case of the computer. It falls apart even more dramatically in the case of the brain. To get the brain to do all the different intelligent things that it does, there has to have been nature's equivalent of engineering. You don't just throw a few billion neurons together and have it do incredible feats like stringing words into meaningful sentences and recognizing faces and calculating the trajectories of moving objects.

We're apt to think there isn't much to pedestrian psychological processes, because they work so well. Just as we're apt to underestimate how complex digestion is until we study the biochemistry of digestion, we're apt to underestimate how complex the mind is from our perspective as commonsense thinkers—exactly because it's designed to work without our conscious awareness. I sometimes think that Gould, as someone who has never been faced with explaining ordinary perception and behavior in his day-to-day work, is apt to underestimate it and therefore to give short shrift to natural selection, which is the only force capable of explaining that kind of complexity.

NICHOLAS HUMPHREY: Some of what Richard Dawkins and Steve Gould go on about in their debate is old-hat, and they ought to stop it. New things have come up since *The Selfish Gene* and since Gould's earlier writing. We're into new territory now. The evolution of evolvability is a question of whether there can be selection for the ability to evolve in changed circumstances. There's increasing evidence that there are ways in which biological systems can be more or less adapted to evolve.

Sex is one very simple example. Sexually reproducing organisms are much better at evolving. There are a lot of other much more interesting levels, much more interesting mechanisms at the biochemical level, where you can get particular sorts of DNA that are better at evolving than others. A lot of the dispute between Gould and Dawkins could be resolved by these new ideas.

BRIAN GOODWIN: Stephen Jay Gould—now, there's a name to conjure with, eh? Stephen has an orientation that I find paradoxical, because the bottom line is that he's a Darwinist. He believes that natural selection is the final arbiter, the final cause in evolution. But for me, natural selection explains very little. Stephen is well aware of this. He talks about morphospace, he agrees that we have to understand morphospace. For me, this is where explanations of form and taxonomy are to be found, and natural selection explains very little.

I have immense respect for Stephen and the range and quality of his ideas, but where we part company is on the matter of emphasis. Stephen believes that biology is a historical science, and natural selection is the final arbiter of what survives and what does not. But that's not the interesting question, which is, What emerges? He's well aware of that. I think he regards me as pushing too much on the problems of emergence and morphology and morphogenesis.

STEVE JONES: Steve Gould is, to put it a bit too flippantly, a snail geneticist gone to the bad. All the worst storms happen in teacups, and the saucers of evolutionary biology have been well and truly filled with metaphorical tea as a result of his views on snails and other things.

Sometimes the message takes a bit of getting at, but it's always worth reading, even if I end up disagreeing with it. In some ways, there's too much baseball in his scientific papers—allegorical baseball, beautifully written speculations based on data which don't, to be brutally frank, support the speculation as well as they might. Ramblings like that fit perfectly well into a popular essay, though. I enjoy, very much, reading some of his evolutionary essays, some of which are masterpieces, there's just no question about it—genuine works of art in the scientific-literary form. But to use that approach in science itself is to be constantly in danger of a triumph of form over content.

GEORGE C. WILLIAMS: I have trouble understanding Gould's persistent efforts to minimize the importance of natural selection, the adaptive changes it produces, and the other things it does. It imposes costs and allows many incidental consequences to arise from the adaptive changes. These have to be related to the adaptations by straightforward cause-effect reasoning. If something happens by chance—for instance, by genetic drift—there immediately arises the question of why drift was stronger than selection in this particular instance.

It's obviously true that there's a lot of chance in evolution, at any level. It's at the higher levels that generally you have sample sizes that are smaller—in the sense that there are not as many species in a genus as there are individuals in a species. In that kind of a situation, the survival of one entity and the extinction of another is much more likely to be a chance event.

The evolutionary process works with whatever it's got. There are no fresh starts; it doesn't design anything new, it just tinkers with what's already there. It may be that what's already there plays some essential role in life, and the life of the organism may turn out incidentally to be useful for something else. If that's important, then it may be subject to modification for that role in addition to its original one. Steve has done a great job of explaining the role of chance in macroevolution and its dependence on historical legacies. There may be a few scientists out there who are as good as Steve Gould, but there are just damn few who are good as he is at writing for a great range of readers.

He, or someone, uses as an example bird wings, which are obviously locomotor appendages. There's a heron that uses its wing to shade the water it's peering into in its search for food, just as we might do with our hand. This is a good example of something perfected as one kind of adaptation happening to be incidentally useful for something else. Whether it will be modified to make it more useful, as an aid to vision, is another matter. What were originally jawbones are now functioning as ear ossicles, which we use for hearing. In this case, they've totally lost the original function and are entirely devoted to the secondary.

This bird-wing example is what Gould calls "exaptation," and it happens all the time. But there's a semantic problem, even in calling the heron's wing a wing. That structure started out as a fin, and just incidentally turned out to be useful for walking on land, and then incidentally that kind of locomotor appendage turned out to be useful for flying with. You simply have to specify your functional perspective. You can say a wing is a flight adaptation, but it's also a flight exaptation, if you are talking about its origin as something used for walking.

DANIEL C. DENNETT: As I look at the history of controversy surrounding evolutionary theory since Darwin, I see a recurring pattern, in which a new wave of theorists comes along, sometimes singly, sometimes in groups, and when they first show up what they think they've got is a refutation of Darwinism; they think they've killed the beast, or at least discovered a major exemption to what they view as the intolerable implications of what the beast says. As John Maynard Smith points out, the early Mendelians—the people early in this century who rediscovered Mendel—at first thought of themselves as anti-Darwinians. They thought of Mendelism as the way to nip Darwin in the bud. They didn't see that in fact it was the salvation of Darwinism. It's roughly half the modern synthesis. In his recent book *Steps Towards Life*, the German chemist and Nobel Laureate Manfred Eigen notes that what he has done is revolutionary, but he knows better: he titles the epilog "Darwin Is Dead; Long Live Darwin." What he acknowledges is that what he has to say is not that revolutionary after all, it's a new wrinkle. It saves Darwin for another day. Stuart Kauffman is the same way. He starts off thinking he's the ultimate anti-Darwinian and he ends up discovering that what he has is a nice improvement to some part of Darwinism.

We'd all like to be considered revolutionaries. Stephen Jay Gould fits into that category. He aspires to bring a certain sort of Darwinism to its knees. He has fought a series of revolutions against what he views as orthodox Darwinism. When the dust clears, however, they aren't revolutions at all. They've made some interesting contributions—some important contributions—but the general public doesn't see that. What it tends to see is Darwinism on its deathbed "as Stephen Jay Gould has shown us." That's just a mistake. That's a major misperception on the part of the public.

What Darwin discovered, I claim, is that evolution is ultimately an algorithmic process—a blind but amazingly effective sorting process that gradually produces all the wonders of nature. This view is reductionist only in the sense that it says there are no miracles. No skyhooks. All the lifting done by evolution over the eons has been done by nonmiraculous, local lifting devices—cranes. Steve still hankers after skyhooks. He's always on the lookout for a skyhook—a phenomenon that's inexplicable from the standpoint of what he calls ultra-Darwinism or hyper-Darwinism. Over the years, the two themes he has most often mentioned are "gradualism" and "pervasive adaptation." He sees these as tied to the idea of progress—the idea that evolution is a process that inexorably makes the world of nature globally and locally *better*, by some uniform measure.

Let's take these three ideas: progress, gradualism, adaptation. I don't offhand know any evolutionist who's ever put them together that way. That's a figment of Steve's imagination. But he tries to keep these three themes always together. If he accuses you of one, the other two are likely to be coming in on the next beat and the beat after that. This is unconstructive, because certainly he would agree that somebody could be, say, a gradualist and not be an adaptationist, or be an adaptationist and not believe in progress, and so forth. In fact, his attacks on all three of these are seriously misguided.

Steve is a gradualist himself; he has to be. He toyed briefly with true nongradualism—the "hopeful monsters" of saltationism. He tried it on, he tried it pretty hard, and when it didn't sell he backed off. There's nothing wrong with gradualism.

Steve, together with Richard Lewontin, wrote a classic, notorious paper on the spandrels of San Marco. It is—supposedly—mainly an attack on "pervasive adaptation," and on the adaptationist program. It completely misfires. Adaptationism is not the bogey they make it out to be, and they don't avoid it themselves. Steve himself is an adaptationist when it suits him.

The question is, do I agree that Richard Dawkins' version of Darwinism—or John Maynard Smith's version—is impoverished? They're the archadaptationists today, and I'd have to say that the impoverishment hasn't been shown to me yet. Certainly Steve hasn't shown it to me in his writing.

Every theme in Steve's trio is good enough in its own limited way. (His more recent business about the importance of mass extinction strikes me as pretty much of a nonstarter.) But none of those themes is original with him; they've been around in evolutionary theory since Darwin. Some people have taken them seriously and some people haven't. None is revolutionary.

Chapter 3

RICHARD DAWKINS

"A Survival Machine"

W. DANIEL HILLIS: *Notions like selfish genes, memes, and extended phenotypes are powerful and exciting. They make me think differently. Unfortunately, I spend a lot of time arguing against people who have overinterpreted these ideas. They're too easily misunderstood as explaining more than they do. So you see, this Dawkins is a dangerous guy. Like Marx. Or Darwin.*

• • •

RICHARD DAWKINS is an evolutionary biologist; reader in the Department of Zoology at Oxford University; Fellow of New College; author of The Selfish Gene *(1976, 2d ed. 1989),* The Extended Phenotype *(1982),* The Blind Watchmaker *(1986), and* River out of Eden *(1995).*

RICHARD DAWKINS: Some time ago, I had a strangely moving experience. I was being interviewed by a Japanese television company, which had hired an English actor and dressed him up as Darwin. During the filming, I opened a door and greeted "Darwin." He and I then entered into a discussion out of time. I presented modern neo-Darwinist ideas and "Darwin" acted astounded, delighted, and surprised. There are indeed indications that Darwin would have been pleased about this modern way of looking at his ideas, because we know he was very troubled by genetics all his life. In Darwin's time, nobody understood genetics, except Mendel, but Darwin never read Mendel; practically nobody read Mendel.

If only Darwin had read Mendel! A gigantic piece of the jigsaw would have clicked into place. Darwin was troubled by the problem of blending inheritance. In his time, it was thought that we were all a kind of mixture of our parents, in the same way you mix black and white paint and get gray paint. It was pointed out that if that was true, as indeed everybody thought it was, then natural selection couldn't work, because the variation would run out. We would all become just a kind of uniform gray.

Darwin struggled and struggled to get around that. Anybody could see that it wasn't true. We didn't become a uniform shade of gray. Grandchildren aren't more uniform in their generation than their grandparents' generation. Mendelian genetics, and the population genetics of the 1930s, was the vital piece that Darwin needed. Darwin would have been delighted and astounded by population genetics, the neo-Darwinism of the 1930s. It's also nice to think that he might have been pleased about kin selection and selfish genes as well.

I approach evolution by taking a "gene's-eye view," not because I'm a geneticist or particularly interested in genetics, but because when I was trying to teach Darwinism, particularly the evolution of animal behavior, I came up against social behavior, parental behavior, mating behavior, which often look as though they are cooperative. It rapidly became clear to me that the most imaginative way of looking at evolution, and the most inspiring way of teaching it, was to say that it's all about the genes. It's the genes that, for their own good, are manipulating the bodies they ride about in. The individual organism is a survival machine for its genes.

I began to develop this rhetoric in 1966, when I was just postdoctoral, and the ethologist Niko Tinbergen asked me to do a course of lectures at Oxford. At that time, W.D. Hamilton's theory of kin selection had just been published, which inspired me. I generalized Hamilton's way of thinking to the whole of social behavior, teaching my students to think about animals as machines carrying their instructions around. The focus of the enterprise was that the individual organisms were tools, levers. Their limbs, fingers, feet were levers of power to propel the genes into the next generation.

I wrote *The Selfish Gene* about ten years later, and that was what I became known for. Many people thought of it as a new idea, which it wasn't. I simply thought that way of looking at things was an imaginative, vivid way of presenting standard Darwinism. It was a new and different way of seeing it.

The idea of the selfish gene is not mine, but I've done the most to sell it, and I've developed the rhetoric of it. The notion is implicit in the approach of the turn-of-the-century biologist August Weismann and in the neo-Darwinian synthesis of the 1930s. The idea was carried forward in the 1960s by W.D. Hamilton (then in London, now my colleague at Oxford) and by George C. Williams, at Stony Brook. My contribution to the idea of the selfish gene was to put rhetoric into it and spell out its implications.

The selfish-gene idea is the idea that the animal is a survival machine for its genes. The animal is a robot that has a brain, eyes, hands, and so on, but it also carries around its own blueprint, its own instructions. This is important, because if the animal gets eaten, if it dies, then the blueprint dies as well. The only genes that get through the generations are the ones that have managed to make their robots avoid getting eaten and succeed in living long enough to reproduce.

Another way of putting it is to say that the world is full of genes that have come down through an unbroken line of successful ancestors, because if they were unsuccessful they wouldn't be ancestors and the genes wouldn't still be here. Every one of our genes has sat successively in our parents, our grandparents, our great-grandparents—every single generation. Every one of our genes, except new mutations, has made it, has been in a successful body. There have been lots of unsuccessful bodies that have never made it, and none of their genes are still with us. The world is full of successful genes, and success means building good survival machines.

The reductionist aspect of the gene-centered view of natural selection makes some people uncomfortable. Reductionism has become a dirty word in certain circles. There's a kind of reductionism which is obviously silly and which no sensible person adopts, and that's what Dan Dennett calls "greedy reductionism." My own version of it is "precipice reductionism." If you take something like a computer: we know that everything a computer does is in principle explicable in terms of electrons moving along wires, or moving along semiconductor pathways. Nobody but a lunatic would attempt to explain what is going on in terms of electrons when you use Microsoft *Word*. To do so would be greedy reductionism. The equivalent of that would be to try to explain Shakespeare's poetry in terms of nerve impulses. You explain things in a hierarchy of levels. In the case of the computer, you explain the top-level software—something like Microsoft *Word*—in terms of software one level down, which would be procedures, subprograms, subroutines, and then you explain how they work in terms of another level down. We would go through the levels of machine codes, and we would then go down from machine codes to the level of semiconductor chips, and then you go down and explain them in terms of physics. This orderly, step-by-step way—what I call step-by-step reductionism, or hierarchical reductionism—is the proper way for science to proceed.

Reductionism is explanation. Everything must be explained reductionistically. But it must be explained hierarchically and in step-by-step reductionism. Greedy reductionism, or precipice reductionism, is to leap from the top of the hierarchy down to the bottom of the hierarchy in one step. That you can't do; you won't explain anything to anybody's satisfaction.

It's a fair point that the gene is an abstraction. At one level, a gene is a bit of DNA. The Caltech biologist Seymour Benzer classified the gene—divided the gene up—and said that we've got to stop talking about "*the* gene." He divided it into the "recon," the unit of recombination; the "muton," the unit of mutation; and the "cistron," which he defined in a particular way, but it approximately amounts to the length of DNA that codes for one polypeptide chain.

So a critic might ask, "Which gene are you talking about?" when you are talking about the gene as a unit of selection. What I've said in *The Selfish Gene* is that I agree that we're not talking about a particular unit. There's a continuum. The only reason why it's important that it's the gene that's the unit of selection is that the gene is what goes on forever. The gene is what goes on for a very large number of generations. Those units of communication that go on through many generations are the successful ones. They're successful by virtue of their effects upon phenotypes. The unit of selection doesn't have to be the cistron. It can be the length of any number of cistrons—technically not one gene, but it would be one gene for my purposes if, say, once it gets together as a cluster it tends to go on for a large number of generations and is therefore available for natural selection to work on.

My key effort would be, if anything, the extended phenotype. The gene is the unit of selection, in that it exerts phenotypic effects. Genes that are successful are the ones that have effects upon bodies. They make bodies have sharp claws for catching prey, for example. If you follow through the logic of what's going on, there's a causal arrow leading from a gene change to a phenotype change. A gene changes, and as a consequence there's a cascade of effects running through embryology. At the end of that cascade of effects, the claws become sharper, and because the claws become sharper, that individual catches more prey. Therefore the genes that made the claws sharper end up in the bodies of more offspring. That's standard Darwinism.

The extended phenotype allows that cascade of causal arrows to reach out beyond the body wall. Extended phenotypes are things like birds' nests, or bower-bird bowers. A peacock has a tail with which it woos females. A male bower bird builds a grass tail, a bower, in the bushes, and dances around it, and that's what attracts the females. That bower made of grass is performing exactly the same role as a peacock's tail. Genes that make for a good bower, a pretty bower, get passed on to the next generation. The bower is a phenotypic effect of genes. It's an extended phenotype.

There are genes for bowers of different shapes. A caddis worm builds a house made of stones. Some might build a house made of sticks, others might build one of dead leaves. This is undoubtedly a Darwinian adaptation. Therefore there must be genes for stone shape, stone color, all the properties of the house; to the extent that they're Darwinian adaptations, they must be genetic effects. These are just examples to illustrate the point that the cascade of causal arrows leading from genes to phenotypes doesn't have to stop at the body wall. It goes beyond the body wall until it hits things like stones, grass.

The extended phenotype is completely logical. It means that anything out there in the world could be a phenotypic effect of my genes. In practice, most of them aren't, but there's no reason in principle why they shouldn't be. Something like a beaver dam causes a flood, which creates a lake, which is to the benefit of the beaver. That lake is an adaptation for the beaver. It's an extended phenotype. There are genes for big lakes, deep lakes: lake phenotypes have genetic causes. You can build up to a vision of causal arrows leading from genes and reaching out and affecting the world at large.

Our genes are like a colony of viruses—socialized viruses, as opposed to anarchic viruses. They're socialized in the sense that they all work together to produce the body and make the body do what's good for all of them. The only reason they do that is that they all are destined to leave the present body and enter the next generation by the same route, sperms or eggs. If they could break out of that route and get to the next generation by being sneezed out and breathed in by the next victim, that's what they would do.

Those are what we call anarchic viruses. Anarchic viruses, the ones that make us sneeze, are the ones that don't agree with each other. They don't care if we die. All they want to do is make us sneeze, or, in the case of the rabies virus, make the dog salivate and bite. But most of our genes are socialized viruses, socialized replicators. They're disciplined and cooperative precisely because they have only one way out of the present body: by sperm or egg.

Before the gene-centered view of natural selection became fashionable, people used to say that if something was good it would happen. This has led some to believe that the adaptationist approach is an easy game. It's been said that you can easily come up with some Darwinian idea to explain anything. As against that, the proper understanding of Darwinism at the gene level severely limits you to a certain kind of explanation. It's not good enough just to say that if something is vaguely advantageous it will evolve. You have to say that it's good for the genes that made it. That automatically wipes out great swathes of possible facile explanations.

Computers are by far the best metaphor for lots of things, because they're so immensely complicated. They resemble living things in so many respects. The whole idea of programming the behavior of a mechanism in advance is vital to the understanding of living organisms. From the selfish-gene point of view, we are robot survival machines, and because genes themselves can't pick things up, catch things, eat things, or run around, they have to do that by proxy; they have to build machines to do it for them. That is us. These machines are programmed in advance.

I've used metaphors like the idea of alien beings from outer space who wish to travel to a distant galaxy and can't, because they can't travel that fast, so what they do is beam instructions at the speed of light, and those instructions make people on some distant planet build a computer, in which the instructions can be run. Instructions are all you need in order to re-create the life-form. It's controlling its programming in advance, given that you cannot program the day-to-day running of the thing. The distant galaxy is too far away: you can't send orders, can't say, "Now do this, now do that," because every instruction takes millions of years to get there. You send a program that anticipates all possible eventualities so that it doesn't need to have instructions sent to it; the instructions are all there. That's what the genes are. Success in evolution is building programs that don't crash. Programs that crash don't perpetuate themselves. The best way to look at an individual animal is as a robot survival machine carrying around its own building program.

I developed the idea of the "cultural meme" as a way of dramatizing the fact that genes aren't everything in the world of Darwinism. The fact that scientists in varied fields have picked up on the metaphor suggests that the idea is itself a good meme. The meme, the unit of cultural inheritance, ties into the idea of the replicator as the fundamental unit of Darwinism. The replicator can be anything that replicates itself and exerts some power over the world to increase or decrease its probability of being replicated. DNA happens to do that remarkably well, but DNA isn't the only thing that in principle could do that. Life on other planets is not going to have DNA but is certainly going to have some kind of replicator. The meme is another example of something that might be doing Darwinism, here on Earth. Maybe we don't have to go to other planets to see another kind of Darwinism going on. Maybe we've got it staring us in the face here, in the form of cultural replicators.

If I represent the ultra-Darwinist view, Brian Goodwin has a much different approach. He thinks he's anti-Darwinian, although he can't be, because he has no alternative explanation. He's primarily interested in embryology—in how you make what is—whereas I'm interested in how what is evolves. He thinks that what's interesting about living forms is almost a special kind of physics. He uses the analogy of a whirlpool, which has a nice spiral shape to it, and the spiral shape comes from the laws of physics. But the laws of physics allow two stable states: either a clockwise spiral or an anticlockwise spiral.

For Goodwin, what genes can do is, in effect, switch the spiral from clockwise to anticlockwise, but they can't do anything else. Everything that's elegant and beautiful about the spiral comes from the laws of physics. He thinks that that's what genes are doing in us—which is easy to believe in something like a snail shell or a ram's horn, because they look like a whirlpool, but Goodwin thinks that's true of everything. He thinks that physics is responsible for the business part of life, and all that genes do is make a choice between the various stable states allowed by physics.

For Goodwin, evolution is just kind of picking its way from one stable state to another. That could be right; it's not contrary to my view, except in detail. There's a continuum between a kind of extreme Goodwin view and my extreme view. I believe that there's not a lot that genes can't achieve in the way of small-scale, gradual, step-by-step change from what's already there. If you are a rhino with a big horn, and if natural selection wanted to change it to a short horn, a sharper horn, a blunter horn, a fatter one, a thinner one—that, to me, is child's play. I'm sure it could be done, whereas Goodwin might feel that only certain shapes of horn are permissible. That's an open question. There could well be serious limitations in what embryology allows. I'm not hostile to that idea; it isn't anti-Darwinian.

The best olive branch I can offer to Goodwin and his colleagues is what I call "kaleidoscopic embryology." Think about a kaleidoscope: you have a little heap of colored chips inside a tube, and the chips are clustered at random, but then you look at them through a series of mirrors, which makes them appear as a beautiful, symmetrical pattern, which may resemble a flower, say. When you tap the side of the kaleidoscope, all that really happens is that the colored chips slip down a bit and change their position, but what you see on the screen is the symmetrical pattern change in elegant ways. Embryologies are kaleidoscopic, in the sense that mutations may produce complicated effects. Embryology itself is a complicated process, such that a random change, a mutation, manifests itself like the image that results when you tap a kaleidoscope. In some cases, the complication is literally a matter of symmetry, if you think about something like a starfish, which has five arms, all the same as each other. Different starfish clearly evolve by ordinary mutational processes. But a mutation that, let's say, changes the shape of all five arms at the same time works in five places at once. Similarly, an earthworm is a long structure with lots of segments that are essentially the same, or a millipede is a long structure with lots of segments. A mutation that makes the legs longer or shorter or blacker or browner works on all the legs at once. Those are two examples of kaleidoscopic embryology. Mutation is filtered through the existing processes of embryology, and the consequences of mutation are complicated. That's what I mean by kaleidoscopic embryology.

Natural selection in the short term favors those mutations that survive, obviously. But there may be a kind of higher-order selection in favor of embryologies that are kaleidoscopic in productive ways. Things like the five-way symmetry of starfish and sea urchins—the embryologies they have may be especially good at evolving. It may be that as evolutionary time goes on, you get not only selection in the short term, in favor of individuals who are good at surviving and reproducing, but every now and again there's a major change in the embryology, which makes it kaleidoscopic in a different way, and which is then favored by a higher-order selection, because certain new embryologies are good at evolving. Perhaps particularly when a continent is cleaned out by a mass extinction and there's a vacuum waiting to be filled, it may be that it will be filled by whatever group of animals has an embryology good at rapidly radiating and evolving into a whole range of new lineages.

Extinctions happen and are enormously important in evolutionary history. There's no doubt that if the dinosaurs had not gone extinct the entire history of life would be different. There probably wouldn't be mammals, for example. Very probably the dinosaurs went extinct sixty-five million years ago for a reason that had absolutely nothing to do with natural selection but because of some catastrophe. That's happened several times in the history of life, and provides the environmental framework in which natural selection works. But only natural selection, short-term selection of short-term advantage, gradualistic change, is responsible for the buildup of complex adaptation. Extinction cleans the slate, allows a new kind of life-form—mammals, in this particular case—to thrive.

My view of this is encapsulated in my phrase "the evolution of evolvability." Certain embryologies may be better at evolving than others. There may be a kind of higher-order selection for life-forms that are not only good at surviving, which is ordinary Darwinism, but are good at evolving. Each time there's an extinction, a new life-form starts to spread and to evolve—in a real sense, to inherit the earth. After the dinosaurs went extinct, the mammals inherited the earth. There may have been something about mammalian embryology which made the mammal body plan good at suddenly evolving, taking advantage of a slate that had been wiped clean. If you wipe a slate clean, there's going to be a mad rush of forms to start evolving to fill all the various traits: carnivore, herbivore, big carnivore, big herbivore, little carnivore, little herbivore, and so on. There may be some embryologies that just aren't very good at radiating out to fill all those vacant slots. There may be others that are very plastic, very good at evolving, very good at taking advantage of changes in the climate and evolving in a widely radiating way.

On the face of it, this idea is rather different from the view I'm associated with. I came to it through playing with my computer biomorphs—the blind-watchmaker computer program. I learned from this program that certain computer algorithms are better at evolving in the biomorphic, blind-watchmaker kind of program than others. I could then imagine a higher-order selection in favor of being good at evolving.

Stephen Jay Gould argues against progress in evolution. We all agree that there's no progress. If we ask ourselves why some major groups go extinct and others don't, why the Burgess Shale fauna no longer exist, I'm sure the answer is "Bad luck." Whoever thought otherwise? There's nothing new about that. On the other hand, the short-term evolution within a group towards improved adaptation—predators having arms races against prey, parasites having arms races against hosts—that *is* progressive, but only for a short time. It's not that everything in evolution has to be progressive, but there will be a period of a million years when a lineage of prey animals is evolving together with a lineage of predator animals, and they're all getting faster and faster, their sense organs are evolving, their eyes are getting sharper, their claws are getting sharper: that's progressive. The prey animals are getting better because their predators are getting better.

I agree that there's no sense in which evolution was ever aiming towards a distant goal of humanity. That would be ludicrous. No serious evolutionist ever thought that. Gould seems to be saying things that are more radical than they really are. He pretends. He sets up windmills to tilt at which aren't serious targets at all.

The "pluralist" view of evolution is a misunderstanding of the distinction I make between replicators and vehicles. Natural selection works at the level of replicators, in the sense that the world becomes filled with successful replicators and empty of unsuccessful replicators. The way those replicators are successful or unsuccessful is by being good at building vehicles, or phenotypic effects. Those vehicles form themselves into a hierarchy of individuals, groups, species, and so on. The differential success of vehicles can be talked about at all levels of that hierarchy. There's a hierarchy in levels of selection as long as you are talking about vehicles. But if you're talking about replicators, there isn't. There's only one replicator we know of, unless you count memes.

Steve doesn't understand this. He keeps going on about hierarchies as though the gene is the bottom level in the hierarchy. The gene has nothing to do with the bottom level in the hierarchy. It's out to one side.

Gould and I aren't just popularizers. Our ideas actually influence and change people's lives—change the way other scientists think, make them think in a different, constructive way. There's a tendency to downplay popularizing. I would not want to use the word "popularizer" for either of us. It's hard to draw a line between the creative and the popularizing. I like to think of myself as a creative force in the field. This differs from reporting—writing a book that explains the existing orthodoxy so that people can understand it. We don't do that. We do something creative: we change people's minds.

On the other hand, when you say we're the two leading evolutionary thinkers, that's not true. The big creative names in evolution today are W.D. Hamilton, John Maynard Smith, and George Williams. Hamilton is the inventor of kin selection. He's now concentrating on sex, because sex is a big problem in evolution theory. What's it for; why is it there? He's provided the latest and probably the most promising theory of what sex is about. He thinks that the reason for sex is as an adaptation against parasites. It's a very exciting, revolutionary way of viewing evolution: evolution as a dynamic, continuous, running-as-hard-as-you-can-to-stay-in-one-place vision. All his career, Hamilton has been original, stimulating, and has inspired generations of research workers to new efforts.

I'm considered by some to be a zealot. This comes partly from a passionate revulsion against fatuous religious prejudices, which I think lead to evil. As far as being a scientist is concerned, my zealotry comes from a deep concern for the truth. I'm extremely hostile towards any sort of obscurantism, pretension. If I think somebody's a fake, if somebody isn't genuinely concerned about what actually is true but is instead doing something for some other motive, if somebody is trying to appear like an intellectual, or trying to appear more profound than he is, or more mysterious than he is, I'm very hostile to that. There's a certain amount of that in religion. The universe is a difficult enough place to understand already without introducing additional mystical mysteriousness that's not actually there. Another point is esthetic: the universe is genuinely mysterious, grand, beautiful, awe inspiring. The kinds of views of the universe which religious people have traditionally embraced have been puny, pathetic, and measly in comparison to the way the universe actually is. The universe presented by organized religions is a poky little medieval universe, and extremely limited.

I'm a Darwinist because I believe the only alternatives are Lamarckism or God, neither of which does the job as an explanatory principle. Life in the universe is either Darwinian or something else not yet thought of.

There's only one general principle in biology, and that, of course, is Darwinism. Nobody doubts the importance of evolutionary theory; nobody doubts that Darwinian evolution is the central theory of biology. But there's a hell of a lot to do in the way of convincing people at large. As you know, 50 percent of the American population don't even believe in evolution, let alone Darwinism. The attacks upon Darwinism, coming as they do from a position of ignorance, tend to build up a reaction. It's undoubtedly true that evolution has happened; to deny that is rather like denying that the world is round. Therefore it's possible for evolutionary biologists to come across as arrogant. Physicists don't have to deal with this.

I'm becoming increasingly interested in computer models and artificial life, because I'm interested in Darwinism as a general phenomenon: what will Darwinism have to be like, in principle, anywhere in the universe. We can't travel to other places where there's life. I believe there probably is life elsewhere in the universe, but we're not sure and we'll almost certainly never know. There are lots of Darwinisms around the universe, but we've got only one to study. We've got lots of animals to study, lots of plants to study, lots of groups of animals and plants to study, but only one Darwinism.

The next best thing to going to another planet is to set up an artificial world, and the computer is the obvious place to set it up in. In the silicon world of a computer, you can pack in such a lot, and there's room for things to go on in that world. You can make your model world have any property you like, and then try to set your Darwinism going in that model world, and with a bit of luck determine which of the essential aspects of this planet's Darwin-ism are essential in the model world and which are incidental, and vice versa.

GEORGE C. WILLIAMS: Although I've criticized it, Dawkins' replicator concept, presented in *The Selfish Gene*, was certainly an important conceptual advance. I have nothing but respect and admiration for Dawkins.

LYNN MARGULIS: Richard Dawkins epitomizes my comments about how scientists rationalize. In his televised response to the Gaia hypothesis, he said, and I quote: "The idea [of Gaia] is not dangerous or distressing except to academic scientists who value the truth." That quote captures the arrogance of Dawkins. I invited him to come and discuss Gaia ideas with Lovelock and me, and he declined even a telephone conversation. I would have happily arranged such a trip and a meaningful idea-tournament with Jim, as Dawkins knew. He prefers to take potshots instead of actually discussing the details of Gaia. When he says Gaia is "dangerous and distressing to scientists who value the truth," he's talking about himself. Gaia is dangerous and distressing to him because, unlike the rest of us, *he* values the truth. The inference of his statement simply exposes his solipsism.

MARVIN MINSKY: I adore Richard Dawkins' conception of memes—that is, structured units of knowledge that are able, more or less, to reproduce themselves by making copies of themselves from one mind to another. A few million years ago, some of our ancestors evolved some brain machinery that was specialized for representing knowledge in a serial and "explicit" fashion, rather than in a parallel and "implicit" manner. These early primate ancestors of ours began to be able to transmit the fruits of their experience by vocal signals—and eventually that led to rapid advances both in already existing abilities to learn and represent knowledge and, perhaps more important, in the social evolution of new ideas. By improving each brain's ability to do serial processing, the entire society was enabled to accumulate knowledge in parallel. Consequently, the very nature of evolution has changed. In the Darwinian scheme, we can evolve only at the level of genes; however, with memes, a system of ideas can evolve by itself, without any biological change. Yet still, we see many of the same phenomena, with evolutionary fitness struggles and all—as when some philosophy evolves a new and convincing argument about why its competitors may be wrong. The interaction of meme propagation with Darwinian evolution has given rise to a new order of things. In particular, it makes possible such phenomena as "group selection" that are less well supported in simpler species. I don't see this much appreciated in the thinking of most other evolutionists, but I and many of my friends consider it an idea of tremendous importance.

BRIAN GOODWIN: Richard Dawkins and I see things in a very different mode, because he's made himself the proponent of Darwinism. For him, Darwin was a revelation. Dawkins was a zoologist, an ethologist, and then suddenly Darwin got to him, and he thought, My God, this is the truth, and everybody should know this truth! He became something of a preacher.

Clearly Richard and I could not be further apart with respect to our perspectives on biology. He's a brilliant exponent of biological reductionism, neo-Darwinian reductionism, down to genes and replicators, and the great thing I find about Richard is that he has made absolutely clear why it is that organisms have disappeared from neo-Darwinism. He thinks he has reached the level of biological reality with genes and replicators. Organisms, as far as he's concerned, are just the packaging for genes; organisms are secondary entities. For me, they're primary, as they were for Darwin. This is where Richard and I have our most passionate disagreements. I see him as the most extreme exponent of what I regard as an unfortunate tendency in biology.

To give a very brief summary of the way he presents neo-Darwinism in *The Selfish Gene* and *The Extended Phenotype*, let me mention four points he makes: (1) Organisms are constructed by groups of genes, whose goal is to leave more copies of themselves; (2) this gives rise to the metaphor of the hereditary material being basically selfish; (3) this intrinsically selfish quality of the hereditary material is reflected in competitive interactions between organisms, which result in survival of fitter variants generated by the more successful genes. (4) Then you get the point that organisms are constantly trying to get better, fitter, and—in a mathematical, geometrical metaphor—always trying to climb peaks in fitness landscapes.

The most interesting point emerged at the end of *The Selfish Gene,* where Richard said that human beings, alone amongst all the species, can escape from their selfish inheritance and become genuinely altruistic, through educational effort. I suddenly realized that this set of four points was a transformation of four very familiar principles of Christian fundamentalism, which go like this; (1) Humanity is born in sin; (2) we have a selfish inheritance; (3) humanity is therefore condemned to a life of conflict and perpetual toil; (4) but there is salvation.

What Richard has done is to make absolutely clear that Darwinism is a kind of transformation of Christian theology. It is a heresy, because Darwin puts the vital force for evolution into matter, but everything else remains much as it was. I suspect that Richard was at one stage fairly religious, and that he then underwent a kind of conversion to Darwinism, and he feels fervently that people ought to embrace this as a way of life.

Where we agree about evolution is with respect to small-scale changes. I entirely concur that adaptation, natural selection, can produce small changes within species—that you can select for different types of dog. But they still remain dogs. The question is, how to cease being a dog and become something else? That's where you need a new principle. Darwinism addresses only small-scale change. It does not address the problem of how you get the large differences of form that emerge during evolution.

There's another sense in which Darwinism is also correct. Natural selection is about stability of different types of life cycle, in various habitats. For a species to survive, to persist, it has to have dynamic stability in a particular habitat. In that sense, Darwinism is trivially true.

But the important point is how the dynamics work, how evolution actually generates these different forms. Richard believes that the accumulation of differences through changing genes gives rise to significant differences, and therefore you can explain species, genera, orders, and families—the whole taxonomic shebang. But there are different categories of form, and this is where the physics and mathematics come in. I believe that there are natural kinds of organisms, so that genera and species are more like the elements in physics, such as hydrogen, oxygen, nitrogen, carbon—distinct forms that are possible. Then there are all the isotopes, a lot of varieties of carbon. But the qualities of carbon remain in the isotopes; you still have the chemistry of carbon. That's what it's like with species: you get a lot of modifications of a species but they're a natural kind. To understand that, you need a theory of biological form, which involves physics and mathematics.

STEVEN PINKER: Reading Richard Dawkins' three books and many articles was a natural turning point in my intellectual development. As a student, I had been convinced by Gould and Lewontin; their view was academically correct in Cambridge, Massachusetts—partly because of the cachet it had for being aligned with left-wing politics, partly because it had come to be seen as the chic and sophisticated position on evolution: you shouldn't be telling just-so stories about the origins of traits. I'd always had some intellectual disquiet about the anti-adaptationist arguments, but I figured I must just have been ignorant about evolution. When I read Dawkins' books, especially *The Extended Phenotype*, I was impressed by the rigorous and subtle exposition of the concept of natural selection and its strengths and weaknesses.

One reason I immediately appreciated the Williams-Dawkins view was that they were trying to explain the aspect of biology that I also deal with—namely, adaptive complexity. I study mental processes that people take for granted because they work so well. When I study the mind, I'm not interested in how Mozart created a symphony, or how Einstein came up with the theory of relativity, those glorious pinnacles of the intellect. I'm interested in the more prosaic things, like walking across a room, asking a simple question, and recognizing a face.

Look at what has to lie behind those abilities. How would you build a machine that recognized a face as easily as we do? Or how would you program a computer to produce and understand ordinary sentences? There has to be an enormous amount of intricate engineering involved. In the case of vision or language, it's software engineering, whereas in the case of the eyeball or the hand it's physical engineering. But still, you need dozens of very delicately arranged subroutines and algorithms to do what we take for granted. What Williams and Dawkins emphasize is the necessity of invoking natural selection to explain improbable, well-engineered traits, and since most of the mind consists of well-engineered, improbable circuitry, the natural conclusion is that anyone who studies the mind is dealing with adaptation: complex products of natural selection that were created because of their ability to solve some problem, a problem that ancestral hominids faced in their everyday existence. If you take the ordinary mind for granted, if you are blasé about ordinary mental activities, then you can afford to think that it might be an accidental by-product of having a big brain, or something that came in through genetic drift, or a lucky mutation. But if you've cut open the frog, as it were, and seen all of the beautifully laid out organs inside, you're much less likely to attribute it all to some fairly purposeless or relatively simple process, like a change in the growth rate of one parameter in embryonic development, and much more likely to look to the kinds of biological forces that can create complex organs.

NILES ELDREDGE: I finally met Richard in 1994. My one previous run-in with him was when he sent me a letter saying he was starting the *Oxford Surveys in Evolutionary Biology* and would I care to contribute a manuscript. I happened to have a manuscript (written with Stan Salthe) that had just been rejected by the journal *Evolution,* with a cover note saying that they didn't publish philosophy or theory but only empirical stuff. I wrote Richard and said, "You aren't going to like this, this is all about hierarchy," and he wrote back and said, "What makes you think I don't like hierarchy?" which was a very amusing and witty thing for him to write, because he's such a gene-oriented, reductionist person. But he does talk about hierarchies, he just handles them differently.

He's been enormously good for the profession. He's funnier than hell. Even when we're being skewered, sometimes I think that what he writes are the best comments against us. But the most clever one was not from Richard but from a geneticist over there, Brian Taylor, who called punctuated equilibrium "evolution by jerks." I had to laugh—even though, of course, it pissed me off.

Richard is not only a communicator, he's also an original thinker. So I have good feelings about Richard Dawkins. Not only that, but a nice aspect of the entire reductionist thing that Williams started in evolutionary biology—the ultra-Darwinians—is that it gave a lot of people a lot of work. Particularly through sociobiology, but the whole gene-oriented thing became a metaphor and also a kitchen industry, and Dawkins is mostly responsible for its promulgation.

W. DANIEL HILLIS: My only complaint about Dawkins is that he explains his ideas too clearly. People who read his books often walk away with an illusion of things being much simpler than they actually are. Just like Marx makes his readers feel that they're suddenly experts with an inside track on history and economics, Dawkins makes his readers feel in a privileged position with respect to biology. This annoys the biologists, especially since Dawkins' ideas are very good. Notions like selfish genes, memes, and extended phenotypes are powerful and exciting. They make me think differently. Unfortunately, I spend a lot of time arguing against people who have overinterpreted these ideas. They're too easily misunderstood as explaining more than they do. So you see, this Dawkins is a dangerous guy. Like Marx. Or Darwin.

STUART KAUFFMAN: Richard and I sit at rather opposite ends of the spectrum. Richard is an inheritor of the pure Darwinian tradition but wants to see selection as the account of everything. He's just ignoring spontaneous order, simply because it's not in the Darwinian tradition, although it's not opposed to Darwinism at all. Richard is a very articulate spokesperson for the most conservative evolutionary interpretation of Darwinism—conservative in the Darwinian sense that everything is due to selection. Darwin himself was a lot more eclectic.

Richard has developed a computer program, described in his book *The Blind Watchmaker*, that generates morphology. I've watched him use it, and it's interesting and fun, and there's less there than meets the eye. I mean to be both accepting and a little bit critical. It's clear that you can generate varieties of morphologies, if you have something called a genotype that makes something called a morphology. You twiddle around with a genotype, and you can mutate out the genotype and generate a different morphology. You can obviously select a lineage of morphologies, and in that sense it's good, because it tunes your intuition about what those branching phylogenies might look like. The part I tend to dislike in what he's done is that there's nothing natural or self-organized or robust about the development mechanisms and morphologies that Richard posits. He simply has computer programs that arbitrarily draw stick figures or whatever. That's not how real development works.

If you look at Richard's blind-watchmaker computer program, what you have, as I understand it, is a computer program that has little subroutines that draw figures on the computer screen that happen to look like animals. In that thing, there is going to be little instruction that says, "Leg-length: do 7 zots." Or "Do X zots." A mutation will say, "Do X minus one, or X plus one zots." Whatever a zot is—for example, the length. That's the developmental mechanism that Richard's working with on the computer. It's a little algorithm, which draws a line. When you mutate the algorithm, you mutate the morphology that the thing draws.

The trouble is that there's a complete arbitrariness about writing a computer program that draws lines on a screen. I mean something specific here by "arbitrariness." For example, in a case of a real morphology: if you take lecithin, or cholesterol, and dump it in water, it forms lipid vesicles; these are bilipid membrane vesicles called liposomes. You get a hollow spherical liposome that looks like a cell membrane. That's a spontaneous thing that lipids do in water. It's a self-assembled, low-energy form. There's a morphology. It's not arbitrary. There's something about the way that physics and chemistry work that gives you something called a bilipid membrane. That bilipid membrane is robust. If you change the kinds of lipids, the temperature, the solute, and the solvent, you'll still get lipid vesicles. You'll get them over a wide range of changes and conditions. There's an inherent robustness to that morphology, which is a fundamental bit of morphology.

You could model that on a computer, but it's not inherently in an arbitrary computer program that draws "morphologies." There's nothing I can see in Richard's approach that indicates he's thinking about explicitly looking at the spontaneously natural forms that can fall out.

DANIEL C. DENNETT: I got to know Richard after reading *The Selfish Gene.* It's sometimes comical, sometimes eerie, how much we think alike. We think so much alike on so many issues that I've come to the opinion—and I joke with him about it—that we should be very careful not to listen to each other too much, because we'll just egg each other on. Whatever our weaknesses are, we share them and tolerate them in each other. We tend to say, in response to each other's work, "Yeah, yeah, right on!" It's very gratifying to find somebody with a different background, different knowledge, and different agendas who so unerringly hits the nail on the head from my point of view!

Some people object to Dawkins as being what I now call a greedy reductionist—that is, they think he's vastly oversimplifying, trying to get the job done with too few levels of explanation. Even though some version of that objection may be true, it's not a big deal. The algorithmic approach as Dawkins presents it is deliberately oversimple. But Dawkins leaves plenty of room for making it more complex. He puts in plenty of warnings that he's giving you an oversimple version of it. The "greedy reductionist" complaint is a tempest in a teapot. Dawkins is not wrong—he's just been too optimistic sometimes.

If you treat the Gould-Dawkins disagreement as essentially one about strategy, it turns out to make sense. It's a constructive way of looking at their feud, and Gould is even somewhat right. Life and evolution—in particular, evolvability—is more complex than Dawkins allowed for. This isn't an earthshaking point, but it's interesting; it might even give Dawkins a nice way of saying, "Thanks, Steve, I needed that."

Maynard Smith has said something very canny about the different attitudes toward nature. There's an urban view of nature and a country view of nature. Gould is certainly expressing the urban view, Dawkins and Maynard Smith the country view. Let's try a little thought experiment. Suppose we brought Aristotle to the present in a time machine and plunked him down on a highway at rush hour. He would of course be amazed. Which would amaze him more? That there were cars and trucks at all? Or that there were so many different varieties of them? Would he say, "Why aren't they all black? Why not all the same shape?" Or would he be more impressed with the fact that there were any of them at all? Presumably both facts are stunning and in need of explanation. Dawkins and Maynard Smith are more impressed with the excellence of the design, the fact that it's possible at all—that there are conditions under which such engineering marvels can come to exist. Gould is more impressed with the fact that the different designs are so different.

The two facts are both important, and they're interdependent, of course. If there were no diversity of design, there wouldn't be much excellence of design. We can well imagine a planet that for billions of years had just a very few simple life-forms and that was it. No diversity to speak of, just a few photosynthesizers, perhaps. After all, that's the way our planet was for most of the time there's been life on it. People like Dawkins and Maynard Smith, who are adaptationists, would be foolish to deny that diversity is an engine that drives optimality. Of course it is! Diversity is an engine that creates what they call arms races, which create design improvements. At the same time, these arms races also create the opportunities for more diversification, so it's a system that feeds on itself. It's pointless to think that there's an opposition between these two ideas. They go together, like ham and eggs.

STEVE JONES: Dawkins has written the best general book about evolution since the Second World War, *The Blind Watchmaker*. It's simple, it misses things—he'd be the first to admit that—but it tells you what it's about and it has a flavor of the subject which no other book's got. I have a great deal of time for it. When Dawkins started, he was a guy who got his feet wet, metaphorically, because he did experiments on the feeding behavior of chickens. There's a general rule that most scientists don't become famous until they stop doing experiments, and Richard illustrates that perfectly.

I would say that he's the most successful popularizer of all. He's been very good for the subject and he's made it accessible to a lot of people who otherwise wouldn't have known about it—people who pay the bills. Also he's been willing to apply brutally simple ideas—the selfish-gene ideas, which are so simple that they ought not to be right—to the most unlikely fields, and they turn out to be right a lot of the time.

In some ways, Richard Dawkins has been the Martin Luther of biology. He's the guy who cut through all the theological mysticism that grew around the true evolutionary church and asked, "What's the big question?" The big questions are the questions you can answer. Any question you can't is by definition tiny and uninteresting.

Chapter 4

BRIAN GOODWIN

"Biology Is Just a Dance"

FRANCISCO VARELA: *Brian should be described as a theoretical biologist. He was introduced into biology from early days, but more recently he has had a structuralist perspective, reaching for fundamental patterns on some expression of life. In that sense, he has come in with a new message, into a biology that's more or less fixated on components and molecules.*

• • •

BRIAN GOODWIN *is a biologist; professor of biology at the Open University, outside London; author of* How the Leopard Changed Its Spots *(1994).*

BRIAN GOODWIN: The "new" biology is biology in the form of an exact science of complex systems concerned with dynamics and emergent order. Then everything in biology changes. Instead of the metaphors of conflict, competition, selfish genes, climbing peaks in fitness landscapes, what you get is evolution as a dance. It has no goal. As Stephen Jay Gould says, it has no purpose, no progress, no sense of direction. It's a dance through morphospace, the space of the forms of organisms.

Will biology join up with physics, take on its flavor, have this notion of rules, organization, regularity, order? The new movement is transforming biology from a historical science, which is what it is at the moment, the objective of Darwinism being to reconstruct the history of life on Earth. Well, that's not the style of physics. Physics is about laws, the principles of organization of matter. We're doing the same thing in biology; we're looking for the principles of organization, the dynamics of the living process. Once that's understood, you're in a position to say, "Ah! History followed such and such a course in expressing and revealing the subtle order in this particular type of organization of matter we call the living state." Thus, the first thing is to understand the living state.

I'm interested in the organization of the living state, which is what theoretical biology is about. How do you define the living state as a dynamic system?

My first contribution was to show that organisms are essentially rhythmic systems accounting for the universality of biological clocks. But I was interested in the spectrum of frequencies showing that control systems oscillate, they have rhythms, the whole organism is an integrated dynamic system that works on many different frequencies. This results in the notion of homeodynamics instead of homeostasis. Instead of having physiological variables that are constant, you have variables that are rhythmic: your temperature, concentrations of substances in the blood, your heartbeat, your respiration, circadian rhythms, menstrual cycles—what is now known as chronobiology. I didn't invent the term, but I gave a strong impetus to the dynamic view of organisms as rhythmically organized entities.

In medicine, chronobiology is regarded as the new wave in treatment of any kind of disease, because you have to be able to tune into the system at the right phase, at the right time. Then there is the notion of dynamic disease. The theoretical biologist Arthur Winfree has developed these ideas—for example, showing why perfectly healthy people suddenly die of heart failure. The reason is that the heart has switched into an alternative dynamic mode—ventricular fibrillation, which is perfectly natural to the heart; it's one of its available dynamic modes. Fibrillation is oscillatory, it's rhythmic, but it just doesn't happen to pump blood very well. You keel over and die of anoxia. The body is a very robust system, all its components interacting with and reinforcing one another. But you can get sudden switches into other states. What may be perfectly natural in terms of the dynamics of the system can be bad news for the person who's experiencing it. Holistic therapies seem to work by keeping the different rhythmic systems tuned to one another.

After working for a number of years on biological rhythms, I switched to the study of biological form. Work on form is going to change the focus of evolution. Instead of being concerned with genes, it's concerned with whole organisms, their transformations, their shapes and forms. We're going to get back to where modern biology started, with Linnaeus and the classification system of different species, relationships of similarity and difference. We'll recover the whole organism as the real entity that's undergoing evolutionary change. It's not the only one; there are ecosystems and other levels. But the organism is absolutely primary. We've lost it and we need to recover it. We need to recover it in medicine, we need to recover it for environmental studies, for ecosystems, for planetary dynamics, for the whole Gaian spectrum of interests. It seems to me that this is where the action is. One of the buzz words I don't particularly like but which has a certain currency is "holism." We're now recovering a holistic view of biological systems.

The small-scale variation and the detailed adaptation of organisms to their habitats are very well explained by neo-Darwinism, but the global problem, the large-scale evolutionary problem, is unsolved. How do you get evolutionary novelty? Emergent order? The difference between squids and fishes and penguins. That's what the science of complexity is beginning to address—to demonstrate how emergent qualities can develop out of complexity, so that you get the emergence of order. The difficulty is making the theoretical work connect with the biological evidence. Most of the modeling currently done on computers is still very abstract, and there's not a lot of detailed evidence as to how that translates into what actually goes on in organisms.

I've always felt that genetics was not going to give me the answers to the problems I was interested in. I know about Dobzhansky, R.A. Fisher, and Ernst Mayr. I went to meetings that the British embryologist, geneticist, and philosopher of science C.H. Waddington organized, and Ernst Mayr was there, so I've discussed these issues with him. I have a great deal of respect for his ideas. He's taken important steps, and made important contributions, but I don't think he's explained the problem that I'm interested in, which is the problem of biological form. It goes back to this original question: How do different types of organisms arise during evolution? That's the question I'm fascinated by. I don't think that any of these people have an answer to it. What they're all talking about is the small-scale, adaptive changes we see in organisms.

This is where Darwin started. He looked at people breeding pigs and cats and dogs and horses. He pointed out that they were selecting on spontaneous variations, producing a great range of forms. Look at the variety of dogs. But they're still dogs. You never go beyond canine characteristics. The question is, How do you get something different? It's generally assumed that if there's an accumulation of enough genetic difference you'll get something qualitatively different. That's a perfectly reasonable hypothesis, but nobody has shown how it works. There seems to be something basic missing. That's what interests me.

At the University of Sussex, I had the good fortune to interact with John Maynard Smith, who had worked with J.B.S. Haldane, one of the founders of the modern synthesis. John was originally a civil engineer. He started his career designing aircraft, but he found that a bit boring, so he became a biologist. He worked with Haldane in London. John says that from casual conversations with Haldane it was clear that he'd thought about the problem that William Hamilton, at Oxford, became famous for, and had provided one of the early solutions to it. That doesn't in any sense discredit Hamilton, it's just that the basic idea had been anticipated by Haldane. Hamilton did not work with Haldane, so he got the idea of kin selection and inclusive fitness quite independently, and he developed it. It was in the air. Hamilton's work is not an area that I follow particularly closely. I don't rank kin selection as a particularly important theory in relation to the problems I'm interested in. Even in relation to problems of organization in social insects, I don't think it's very important.

It's often argued that the reason social insects—ants, or bees, or termites, or whatever—are all so cooperative is that they're all related. They all share the same genes, so they cooperate. If you don't belong to the same family, you're not going to cooperate. That's the basic idea behind Hamilton's notion of inclusive fitness. This isn't a satisfactory explanation of cooperative behavior, because it doesn't show you how the phenomenon arises. It's the same sort of proposition as the genetic argument about form.

People assume that because genes can alter form, therefore they cause it. But they don't actually explain how the form comes into being. Similarly with the patterns in social insects: you have to go further than just to say that they're related to one another and so they cooperate. That's a bit of a trivialization of Hamilton's thesis, but nobody from the Oxford group, as far as I'm aware, has ever demonstrated how the actual phenomena of cooperation emerge from the dynamics of group interaction.

George Williams is very important, because of his work on the evolution of sex. Sex is a big problem for neo-Darwinists, because organisms that reproduce without sex—such as strawberries making new plants from runners—are much more cost-effective than having two plants, or two organisms, that need to come together to make one of their progeny. Why have sex, when it's more efficient to be without it?

Williams has very sophisticated arguments on how this came

about, and what the advantages are in terms of diversity and variation—mixing the genes in populations. It's the genetic algorithm. Organisms are more effective in exploring the potential space of genes if they mix different genomes together. Dawkins was strongly influenced by Williams' arguments, as was Maynard Smith. Within the terms of neo-Darwinian axioms, Williams gives plausible arguments.

Niles Eldredge and Stephen Jay Gould brought people's attention back to the problems of large-scale changes in evolution. How do you get new species? Their notion of punctuated equilibrium addresses a real problem here. Eldredge and Gould looked at the fossil record. What happens? It's absolutely startling! You don't get one species turning into another. A species emerges, it lasts for several million years, and it disappears. Five hundred million years for some of these species, a few million for others. Species emerge suddenly, not slowly. Punctuated equilibrium keeps these problems of emergence in focus. Of course, Eldredge and Gould got a lot of flak; they were accused of being Marxists. They were talking about big changes that happened, biological revolutions. Eldredge is no Marxist, but Gould has a Marxist background. More power to him. He was accused of actually smuggling revolutionary doctrine into biology. Absolute crap! What he was doing was looking at the evidence.

In Britain, the cladists, who construct taxonomies by detailed computer studies of character distribution in species, faced similar criticism. They used the following argument: to understand the relationships of similarity and differences between organisms, you must use strictly logical criteria that are independent of history. That was regarded as heresy, because it was considered that the whole of taxonomy, of classification, was based on history, on descent with modification. The cladists were accused of abandoning Darwinism, just as Gould and Eldredge were accused of abandoning the fundamental principles of Darwinism based on the accumulation of small adaptations. But Darwinism itself fails to explain evolutionary novelty.

There's also some misunderstanding of the role of physics in biology. With a focus on genes and how they change in time, there's a tendency to ignore the spatial dimension of organisms. But organisms are spatially organized systems, and to describe their spatial patterns you need field theories like those used in physics to explain spatial order. In a developing organism, these are morphogenetic fields, the fields that generate form during embryonic development from the egg to the adult. Biologists do encounter this concept of field, but it isn't developed. One of the reasons for this is that a biological education includes very little physics and mathematics, which are necessary to understand how complex forms can arise from initially simple beginnings, such as a fertilized egg. There are laws—principles of morphogenesis—involved here.

There's a difference between these laws of form and principles of engineering. Principles of engineering are structural principles; they don't tell you how things spontaneously change and develop. They tell you how to put things together so they'll have certain properties. What you need in order to understand how form emerges in a developing organism is something much more like the physical theory of the origin of the cosmos—something like Hawking's ideas. Or the origins of the planetary system, where you start with a mass of gas, and out of that you gradually get condensation of the planets orbiting the sun. That's a real evolutionary problem of form. How does the elliptical form of the planetary orbits emerge? That's the kind of problem you deal with in developing organisms, except that living organisms are much more complex than planetary systems.

To understand morphogenesis you need field theories that deal with relationships of processes and structures in time and space, and how these can change. That's why, for me, physics is absolutely fundamental. But it took me a long time to understand what was required, and I didn't do it by myself. There's an old tradition of doing biology this way. This approach can now take off with the power of computers, because these fields in biology are mathematically very complex.

To introduce the problem: I like to compare morphogenesis with hydrodynamics. Suppose you have a fluid, and you want to understand why it takes certain shapes and forms: wind passes over it and it goes into waves, or you get whirlpools at the bottom of waterfalls. Why do liquids take these forms? What you need is a physical theory of fluids, which are a state of organization of matter. It's the same type of problem with organisms. Organisms are states of organization of matter. There are certain principles of spatial order in organisms, in cells, in the way cells interact with one another, and these can be written down as rules or equations, and then you can solve the equations on a computer and find out what shapes emerge, exactly the same way you can with liquids.

The hypothesis here is that life is a particular state of organization, a physical and chemical system. The problem is to find out what the rules are that apply to this state of organization of living systems.

So I see myself more as a physicist than an engineer, involved in a new synthesis of physics and biology. It's been attempted before, most notably by the Scottish zoologist D'Arcy Thompson, in his book *On Growth and Form*, in 1917—an amazing achievement. He single-handedly defined the problem of biological form in mathematical terms. It's changed now, because we have new mathematical tools and a lot of new knowledge about organisms.

The metaphors I use are related to emergence and creativity and the concept of a creative cosmos. Evolution is an aspect of this creativity. Alfred North Whitehead was a wonderful philosopher of process and creativity. The central metaphor I feel is emerging in the new biology is all connected with creativity. You see in genetic reductionism Whitehead's fallacy of misplaced concreteness, par excellence. Genes are not themselves creative but function within the context of the organism, which is.

Whitehead's phrase for evolution is "the creative advance into novelty." This dance of creation is a never-ending dance that goes nowhere but is simply expressing itself. In the postmodern age, we can let progress go and talk about process as a creative dance. That's what evolution is about. Evolution has no point, no meaning, and no direction. It's just itself. Gould celebrates this in *Wonderful Life*. He goes over the top in certain ways, but the basic message is a celebration of the unbridled creativity of life.

As I see it, each species has its own nature, its own characteristics. What organisms are doing is expressing a particular type of order and organization that's deeply within their own beings. All organisms are basically equivalent, because we're all part of the same process, as Darwin described. What doesn't come out clearly in Darwinism is the notion that what happens in evolution is that organisms express their own natures, so that they are to be valued for their being rather than for their function.

Darwinism stresses conflict and competition; that doesn't square with the evidence. A lot of organisms that survive are in no sense superior to those that have gone extinct. It's not a question of being "better than"; it's simply a matter of finding a place where you can be yourself. That's what evolution is about. That's why you can see it as a dance. It's not going anywhere, it's simply exploring a space of possibilities.

There's a focus on competition in Darwinism because of the notions of progress and struggle. Now we get into theology and how it influences Darwinism, through the Calvinist view that people who have the greater accumulation of goods have proved themselves superior in the race of life. That for me is a whole lot of garbage that can be chucked. Once you get rid of it, you're into a different set of metaphors, related to creativity, novelty for its own sake, doing what comes naturally. Instead of the image of organisms struggling up peaks in a fitness landscape, doing "better than"—which is a very Calvinist work ethic—there is the image of a creative dance.

There's still struggle, in the sense that if you're going to be creative you have to believe in your ideas and struggle for them. Every single species has a struggle. But because there is as much cooperation among species as there is competition, the struggle is to express your being, your nature. These are metaphors whereby science can begin to connect with the arts: people being creative and playing. There's nothing trivial about play. Play is the most fundamental of all human activities, and culture can be seen as play.

There's too much work in our culture, and there's too much accumulation of goods. The whole capitalist trip is an awful treadmill that's extremely destructive. It needs to be balanced out. This is why indigenous cultures are beginning to be recognized for their values—because they were not accumulating goods; they were living in harmony. They were expressing their own natures, as cultures. Nature and culture then come together. This is what I refer to as the science of qualities instead of a science of quantities—that is, accumulating things, accumulating genes, accumulating gene products, balancing out your costs and benefits, always trying to accumulate more. Instead of those images, we have images of qualities, which include esthetics, relationships, creativity, health, and quality of life.

These conclusions are the result of attempting the unification of biology, play, and mathematics. Mathematics is a tool for exploring what constitutes the nature of something. If you're interested in nature, mathematics is terribly good at uncovering the nature of something in a particular form. But it's a third-person or "objective" perspective, whereas the first-person, experiential component is what goes with play. Mathematics is a tool for exploring generic forms, natural forms, and a way of looking at their stability and their dynamics and their change, and so on. But you have to couple that with the internal, experiential aspect of creativity. This is what the postmodern science of qualities is about.

I understand "postmodern" in this context to mean that you don't have competing paradigms, you simply have different paradigms. In postmodern science, you have alternative paradigms, and you have a sense of values. Depending on what you want to do in the world, you'll choose one paradigm or another. Therefore values come into the choice of paradigms—values determined by what your goals are.

There are many other qualities that go with postmodernism. I stress the ones most germane to a science of qualities—a way of looking at biology in which you recognize the intrinsic value of organisms. And that connects with environmental action, by which you respect other organisms, other species. This leads in the direction of valuing biodiversity, preservation of the environment, validation of indigenous cultures and their ways of doing agriculture, instead of the monoculture mentality that goes with our agricultural system. Monoculture is at variance with indigenous agriculture, which preserves biodiversity and is more productive, because, given a fluctuating environment, diversity copes with the variations from year to year.

This shift of values in biology represents an alternative to neo-Darwinism. I don't want to eliminate neo-Darwinism—the view that evolution occurs by random genetic variation and natural selection of the superior variants. Competition, climbing fitness peaks in fitness landscapes, monoculture—it's always a notion of what's the best, of designing the best species. If you want neo-Darwinism, there it is. You use it. I don't like it. I like the alternative that's emerging in a science of qualities.

Stuart Kauffman went to Santa Fe, and he invited me to visit. I met Doyne Farmer, and I met a bunch of people at Los Alamos. This was just before the Santa Fe Institute was established. It was being talked about, but it didn't have a site. Then I visited during the institute's first year, and I thought, This is a fantastic idea! Stuart invited me to serve on the science board, and I was very pleased to do so. From then on, I've visited Santa Fe once or twice a year and taken part in that enterprise. It was a brilliant vision. It was Murray Gell-

Mann's vision, and the chemist George Cowan was the guy who implemented it—the guy on the ground. George was great, because he combined a visionary perspective with a practical orientation; he knew how to get there. It's fortunate that so many different types of individuals come together in that Santa Fe enterprise. I've never seen anything succeed so quickly in my life.

What I find remarkable is that the new paradigm is both mathematically more rigorous and fits the phenomena of biology better than neo-Darwinism, which leaves out development and organisms. We now have mathematical models that allow us to show how development occurs. Everybody acknowledges that evolution must include the evolution of development, because you don't get organisms without their development. When you put that into evolution, the whole scene changes. You get a shift of perspective, because organisms become real entities again, living in their own space, so you suddenly recognize them as equivalent beings to yourself. Not just because we're all the results of the same evolutionary process but because of their intrinsic values. The result is that you value nature the way you value works of art.

MURRAY GELL-MANN: Brian Goodwin is especially interested in developmental biology. He's curious about the limitations that physicochemical laws place on how biological systems can operate. Now it's evident that when biological evolution—based on largely random variation in genetic material and on natural selection—operates on the structure of actual organisms, it does so subject to the laws of physical science, which place crucial limitations on how living things can be constructed. But Brian, in stressing the importance of that subject, also implies somehow that work on the informational aspects of evolution and other complex adaptive systems is not particularly interesting. I find that a little odd. Perhaps he doesn't really believe what he says and is just being mischievous.

STEPHEN JAY GOULD: Brian's main commitment, contrary to the norm in twentieth-century England, is to represent one of these great traditions of Western thought, structuralism in biology, which is a nonselectionist, nonhistoricist view. It's basically the argument that laws of form and structure of matter constrain very much how organisms are built, and therefore the main features of organic design are neither necessarily specific adaptations built by selection (which is functionalism, the opposite view) nor are they historical contingencies (as I would argue in many cases) but they are representations of inherent natural patterns.

The major statement of that line of thought in English writing is D'Arcy Thompson's *On Growth and Form*. It's a grand tradition. I don't accept it to the extent that Brian does, because of my own commitment to historical contingency. But I'm very interested in structuralism. It's a way of thinking about form which works in many cases. It's another way of critiquing the pure functionalism of the adaptationist program.

STEVE JONES: I've read some of Brian Goodwin's stuff, and I find it extremely hard to follow. That could be because I'm stupid. But I did embryology, I did development, I read molecular biology; that isn't so difficult to follow. Goodwin makes it hard. It's not an approach I like. I think he's a mystic. Anybody who goes to Santa Fe—there's something in the air there that's catching. Complexity is catching, that's the trouble.

RICHARD DAWKINS: When Brian Goodwin was at Sussex, associating with John Maynard Smith, we thought he was doing a rather good thing of just being a bit off the wall, because Maynard Smith is so sensible. We thought when he moved to the Open University, where they're all crazy, then Brian Goodwin would switch back and become sensible. I don't think that's happened. He has an interesting point of view, which I have argued with in print. It's a genuinely interesting idea that the range of variation that natural selection has available to it is not the continuously varying or possible variations that some extremists' presentations of Darwinism might have us believe; that somehow almost any change in the existing morphology, providing it's a small, gradual change in some quantitative variable, must be possible.

It's a genuinely interesting possibility that the underlying laws of morphology allow only a certain limited range of shapes. I don't think there's much good evidence to support it, but it's important that somebody like Brian Goodwin is saying that kind of thing, because it provides the other extreme, and the truth probably lies somewhere between the Goodwin extreme and the hypothetical extreme, and it's possible for sensible people to slide somewhere along that continuum.

NICHOLAS HUMPHREY: I didn't get to know Brian until recently. I heard him give a lecture at a Waddington conference, and I was amazed. I thought, What is this man on about? He was proposing a theory to replace natural selection as the motive force in evolution. His argument was that we don't need natural selection. All the beautiful structures we see in the world are just there for the taking; they just emerge out of complex dynamical systems, because the world is full of "attractors." I thought, He can't be serious. Maybe these things are mathematically possible, but he can't really mean that no other organizing principle is needed. We had a big argument, and he did mean just that.

I was not only surprised but quite upset to hear what Goodwin was up to. But then I went away and read some of Stuart Kauffman's stuff and talked to the mathematician Ian Stewart, who was also at the conference, and I realized that there was a revolution taking place that I knew nothing about.

There's no question that the idea of attractors might in some circumstances provide a simpler account of how biological structures have come into existence than natural selection does. But the trouble with it is that it's very hard to criticize. You can always say, after the event, that there must have been an attractor for whatever has occurred—because if the world ends up in a particular configuration, then it must have been attracted to it!

After hearing Ian Stewart and his friend the reproductive biologist Jack Cohen brainstorming, I got the impression that they're prepared to posit attractors for absolutely anything. For example, the human mind has a short-term memory capacity of about seven items—the "magic number 7." I asked Ian and Jack whether this might suggest an attractor for the number 7 at work in the brain. Why not? they said. Or for another example, we have five senses, five qualitatively distinct ways of experiencing the world. Does each of these sensory modalities correspond to an attractor, and are there are only five such attractors possible? Why not? they said. It made me think that this is an easy game. We can find attractors to account for any stable patterns we find.

DANIEL C. DENNETT: Since I haven't met him, I view Brian Goodwin as an archetype, an individual without peculiarities. I see him as the standard-bearer for a certain position in logical space. It's not surprising that it's occupied, but it seems to me to be mainly wrong.

Brian Goodwin is a romantic, who wants to deny that biology is ultimately engineering. You can get the idea by considering a parody. (Like all parody, it's surely somewhat unfair, I grant.) I don't think the following position is occupied, but maybe it is: Maybe somebody somewhere in the engineering world has said, "You people think that engineers have carefully *designed* all these artifacts—automobiles, television sets, and so forth—but in fact they've only *discovered* the few possible artifacts that there are. These aren't in any interesting sense designed; the laws of physics let there be television sets and let there be cars, but there's much less scope for alternatives than you'd think." Goodwin's position is the analog of this position applied to biological artifacts—that is, organisms.

I don't think there's anybody who thinks there are deep, fundamental laws of automotive engineering, but when Goodwin says that there are laws of form in biology, he's making what I view as an equally implausible claim. All the regularities of biology strike me as being exactly like the regularities of engineering. Thus cars have their steering wheels placed so that the driver can see forward, not backward. It's a deep regularity of automobiles. It's not a law of nature, it's just that it would be stupid to do it any other way. Locomoting organisms tend to have the eyes and mouth at the front end, for much the same reason. It's not a deep law of nature; that's the way to design something that has to fend for itself. Goodwin is unhappy with that style of explanation, and would like to see deeper laws of physics explaining all this. It's an idea I think I do understand—and don't believe at all.

LYNN MARGULIS: Brian Goodwin seriously and appropriately criticizes neo-Darwinist thinking and the voguish ideas of biology today. He's an exactingly appropriate critic, which I admire. He also poses interesting and important problems in developmental biology. Unfortunately, he can't possibly approach the answers, because the answers lie in a domain that he doesn't study. That domain is microbiology, including comparative metabolic chemistry and knowledge of the microbial communities which bacteria and protoctists compose. I doubt if he realizes that answers to the compelling questions he poses exist elsewhere.

FRANCISCO VARELA: Brian should be described as a theoretical biologist. He was introduced into biology from early days, but more recently he has had a structuralist perspective, reaching for fundamental patterns on some expression of life. In that sense, he has come in with a new message, into a biology that's more or less fixated on components and molecules.

Brian and I have some important differences, but where we come together is in a search. Brian's inspiration is a search for a way to talk about the organism as a unitary structure. He's working more with embryological patterns—"basic morphs," as he calls them. This is a very royal approach—and in that sense I like it very much—looking for fundamental patterns and forms, which is related to René Thom's catastrophe theory.

Chapter 5

STEVE JONES

"Why Is There So Much Genetic Diversity?"

STEPHEN JAY GOULD: *I like Steve Jones' work. I've read most of his scientific papers. I work on pulmonate snails, and he's one of the best in this little field. I don't know him very well. He's a very good scientist. He's followed the path of a media person, but in my professional world—snail biology—his science is very good.*

• • •

STEVE JONES *is a biologist; professor of genetics at the Galton Laboratory of University College London; author of* The Language of the Genes: Biology, History, and the Evolutionary Future *(1993); coeditor (with Robert Martin and David Pilbeam) of* The Cambridge Encyclopedia of Human Evolution *(1992).*

STEVE JONES: I've spent a long time working on snails. This seems an odd thing to do, but they remind us of one question that remains unanswered, and is effectively forgotten by many biologists: Why is there so much diversity? Without it there could be no genetics, no evolution, and—probably—no biology at all. It's a question that periodically comes to the surface and then sinks again; just like a political issue, it appears, disappears, and remains unresolved. We have the beginnings of an answer as to why, in some places, one snail species is so variable, but we have no real idea why in any species anywhere at any time no two individuals are identical. That's an essential question of evolution. All others flow from that.

I'm an evolutionary biologist. For most of us, from Darwin on, the organism you work on is the second question. The first is, What is going on in evolution? Inevitably, though, one ends up talking about the particular rather than the general. Snails are a microcosm of what Victorians used to think evolution was about, in that they're strikingly different from one another in appearance, and often from one species to another. Perhaps they're a good particular case to study for that reason alone—and, of course, they're easy to catch.

I don't think any scientists can explain, with any honesty, how they got into their own particular topic. My theory is that most people do it by accident, in spite of what they claim later. I got into evolution by chance. My tutor was a chap called Bryan Clarke, a very able scientist who worked on the genetics of snails. In those days before molecular genetics, snails were one of the few creatures whose diversity was easy to study, so it was less odd than it might seem today. I was assigned to his tutorial group on the basis of the first letter of my surname, and inevitably I started working on the same thing. Fortunately, it turned out to be fascinating (to me, if to no one else) and it's still my prime scientific interest—although I can't be accused of being a narrow specialist, as I have now moved into slugs.

Doing molluscs was probably the worst career move I ever made, because the people around me are now household names in biology—people like Ed Southern, who invented the Southern blot, a central technique in the new genetics. I could easily have gone into that field and perhaps have become slightly less obscure than I am today. But I didn't, and I don't really regret it.

Steve Gould has concentrated a lot on the differences in shape and form in the Bahamian snail he works on—a bigger and more difficult question. I work more on differences in color and pattern—and increasingly I'm asking the same questions about diversity at the DNA level. The shell-pattern variation is a classic of evolutionary biology. It was first looked at in the nineteenth century. We now have data on well over a million individuals who have been scored for their physical appearance—an awful lot of information about differences.

To summarize a lot of work, it seems clear that as you go across Europe, the snails' genes are different for reasons having to do with thermal relations in sunshine. Dark objects—and genetically dark snails—heat up more in the sun, and those genes are rarer in the south. The same is true on the smaller scale of a few miles. I did a lot of work on that on the border of Bosnia and Croatia. A few months ago on the news, I saw the town I was based in burning to the ground.

That leaves the more difficult question: given that in different places individuals with different genotypes are favored, why is every snail in a particular population not always the same?

We have, I think, sorted that out. What I did was to use the snails themselves as ecological monitors. I developed a paint that fades at a known rate when exposed to the sun. Put small spots of that on the shells of snails of different genotypes, return them to the wild, and come back after a month; the paint indicates the amount of solar energy soaked up. There are big differences between animals of different genetic constitution. They choose different times of day, or different parts of the habitat, to be active in, suggesting that an ecologically complex habitat might support more genetic diversity.

I developed another technique to test this, which has become known as "Jones' balls." These are snail-size spheres made of plastic, which you throw into a habitat. Then you pretend you're the sun, to put it childishly, and you take the world's cheapest satellite, which is a ladder, and scan the habitat on a track from sunrise to sunset. This measures the pattern in which snail-size objects are exposed to the sun or hidden by the vegetation at different times of day. It gives a kind of snail's-eye view of the universe. I suppose it's a trivially simple idea, but it works. It measures habitat diversity as perceived by the snails. There's a good fit between genetic diversity, ecological diversity, and individual choice of microhabitat.

It's a fairly new idea to try to get a view of ecology through your organism's eyes, as it were. We've succeeded in doing the same kind of thing with fruit flies. The whole of genetics used to be *Drosophila,* and it's still a very important organism, because so much is known about it in the lab. Linda Partridge, a colleague of mine, and I, and Jerry Coyne from the University of Chicago were interested in trying to assess the ecological relationships of *Drosophila* using genetic means. We used a mutation that was temperature-sensitive: eye color depended on the temperature the fly experienced during development. We did the experiment in Maryland, where it was hot and steamy. We released millions of flies containing this mutation and collected their offspring, which had developed in the wild.

The flies themselves were acting as living thermometers. We could tell from their eye color what temperature they'd grown at. They occupied an extraordinarily wide range. Because flies growing at different temperatures emerge at different sizes, this explained an awful lot of the variation in shape and size which previously had been thought to be genetic. And, in turn, that helps explain another large but largely ignored question in evolution. For most creatures, it pays to be big. It makes you a better mate, better at dealing with enemies, better at coping with heat and cold. Why, then, is there any variation in size? Perhaps it's because most of the differences—in fruit flies, at least—are environmental, and not genetic at all.

I also have a vicarious interest in sex. After all, it's just a machine for generating diversity—differences between parents and offspring. Nobody really knows why sexual reproduction is there. About the only way to study it is to look at those few creatures who've given it up. Most slugs are hermaphrodites, but nearly all are relatively decorous about it; boy-girl meets girl-boy and nature takes its course. Some, though, have taken the easy way out. They fertilize themselves, effectively abandoning sex altogether. Their genes show that these species are essentially a mass of identical twins, with no diversity at all. We don't yet know why they do this: the only real pattern is that it pays to give up sex in the cold—in Norway, as compared with Spain, for example.

The pivotal influences on my work have been Bryan Clarke and Dick Lewontin. When I finished my Ph.D., I wrote to Lewontin in Chicago, asking him for advice as to where I could go for a postdoctoral fellowship. He wrote back almost by return mail saying, "Thank you for your application, which is accepted. We expect you at the beginning of next month." I hadn't even applied, but I went there like a shot and learned, more than anything else, how little I really knew.

He was then approaching the question I have been studying ever since, which is why genetic diversity exists. Dick is a man of tremendous brio and enthusiasm, who has the ability to fire people up with his ideas, however good, bad, or indifferent they might be. I have to say that I was greatly enthusiastic about one of his bad ideas—that it might be possible to take an isolated population of fruit flies living out in the California desert and use it as an artificial laboratory, change the flies' genes by flooding them with genetically different flies and see what happened to their evolution over a couple of years.

That was a great time. I traveled all over the deserts of California, into Mexico, looking for isolated populations. After three years and a lot of money and a deep tan, what we basically found was that these populations weren't isolated at all; there were flies flying in and out all the time. In its own narrow way, that was interesting for *Drosophila* genetics—though it certainly wasn't going to change the course of evolutionary biology.

Lewontin excited me about science more than anybody else has ever done. He did the same for lots of people. If you trace the family tree of evolutionary biologists in the world, a suspiciously large number of them lead straight back to him. He has been pivotal in the subject.

He's sometimes a pernicious influence, though, in the sense that Marx or St. Augustine were. They may both have been wrong, but life would have been a lot less interesting if they hadn't been around. At least they forced people to think about their ideas. Dick is an evolutionary gadfly, attacking whatever the dogma of the day might be. He's the embodiment of the idea that science is the art of the disprovable. He's destroyed lots of ideas, and that's a useful thing to do. He does it superbly, but science needs more than iconoclasts. It needs some people—hacks, like me—to build the icons up, even if their fate is to be knocked down by the Lewontins of this world. Still, I wish there were more people like him around.

I do know a lot about snail genetics. It's my narrow, limited, unintellectual kind of field. In many ways, though, it's a microcosm of evolutionary biology at its worst. Its literature is filled with the great vaguenesses of evolution—with words that, when you deconstruct them, are like shoveling fog; they don't mean much. "Coadaptation," "adaptive landscape," "punctuated equilibrium"—what I sometimes think of as theological population genetics. They're words that don't help at all when you're trying to decide what experiment to do next.

Words like these reflect the view that somehow one gene is there because it has adapted to the other genes that were there already. That the world somehow is a beautifully harmonious structure is an optimist's point of view: everything fits beautifully together, and if you see the whole edifice you don't have to worry about how it's constructed, it just stands up.

That's a pernicious idea. It's an anti-intellectual, working-out-God's-plan, know-nothing kind of idea. In what must have been a moment of extreme tedium, I once read a book by a South African general, Jan Smuts, called *Holism*. Smuts was a strange, interesting guy, who dabbled in philosophy. Everything you saw in the world was all part of a great scheme, and there was no point in trying to work out what individual parts of the scheme were for, because it made sense only when you saw it as a whole. He was a rather weak philosopher. But his idea pervades a lot of biological thinking. Evolution is a magical thing, with an intrinsic beauty of its own, which you can't hope to break down into the individual genes that make it happen. In other words, there's a limit to reductionism.

Well, maybe there is; but the beauty of reductionism is that it gives you something to do next. Once you start saying that something's unexplainable, then there's no point in trying to explain it. Steve Gould and Dick Lewontin made a famous and very funny attack on reductionism; but in some ways it shows the weakness of what I guess we can call the Argument from Smuts. It was the talk on the spandrels of San Marcos, at the Royal Society. It made an important point about hyperadaptationist views—that everything is the way it is for a reason that can be explained in simple biological terms. The extreme reductionist might write learned books about the Spandrel School—about the deep artistic reasons why the painter made his paintings in this particular shape, and what he was trying to say by not making them square. But the shape of the paintings is there for a reason that had nothing to do with painting. Gould and Lewontin made great play with the parallels between the Spandrel School and the many evolutionists who say that every character in every animal is there for an adaptive reason and if you look hard enough you'll find it.

There's some truth in their argument, but to accept it as the only truth is basically to give up and walk away, to stop being an ornithologist and turn into a bird-watcher. You become somebody who observes rather than analyzes. What they're saying to lots of biologists is, "Abandon hope, go home, and become a liberal-arts graduate!" I may be overcriticizing the Lewontin and Gould view; both of them like to poke people with their sharp pitchforks. The spandrels were a particularly successful poke. But what happened as a result of the famous spandrel paper? The answer is, not much.

Contrast that with the views of someone who is definitely not on the side of the Angels of San Marcos, Richard Dawkins. His views are—to simplify them—simplistic. You can deconstruct everything down to a series of units, the genes—although Dawkins himself would admit that it's naive to say that organisms are just vehicles for carrying DNA around and that everything they do is in the interests of their "selfish genes." But his metaphor has turned out to be extraordinarily productive and useful, because it gives you all kinds of ideas about how to test it. Again, reductionism provides the scientist with raw material, which is a lot more than spandrelism does. That's the beauty of the selfish-gene idea. You can grab it and test it. You can look at the idea and look at the genes. It may well be that the idea will turn out to be wrong. But it sparked a lot of very interesting work. The idea nowadays is that the most fundamental rules of biology—Mendel's laws themselves, even—are a reflection of a truce in a battle between selfish genes. That's a remarkably interesting thought, which leads to some testable predictions.

If I learned anything from my work on snails, it's that reality is getting your feet wet. The only way to approach the truth in snails, or in anything else, is to go out and do the work. Leave pontificating to the Pope. That may sound trivial, but it's important. Science is data-led, not theory-led. I never feel usefully employed in science except when I'm gathering data. Unfortunately, the system conspires to stop you; and instead I give interviews like this or write brittle little pieces in *The Daily Telegraph.* Old age, idleness, administration, grant-starvation, all those terrible things don't help either.

The questions I ask myself today are the questions I was asking thirty years ago. The only thing that's changed in the last thirty years in genetics is that humans have become the new fruit flies—the organisms that are technically accessible to asking questions about genes. How different are two people? Why are two human groups different from each other? What's the history of human diversity? That's what I shifted to, but now I'm increasingly a voyeur of science rather than a doer of science.

Because I do a lot of writing and broadcasting, I'm better known as a geneticist by the general public than I am by other geneticists. Although I write a lot about it, I've never done any serious work of my own in human genetics, so I'm a spectator of the subject rather than a participant. I'm grateful to my colleagues for being slightly less cynical about that than they have the right to be. I think they see that there's a role for the reporter in science. However, I can console myself with the thought that I'm one of the top six snail geneticists in the world, out of a field of perhaps half a dozen.

RICHARD DAWKINS: I've enjoyed Steve Jones' recent book *The Language of the Genes*. He's a little bit too eager to bend over backwards to be politically respectable, because of the unsavory history of genetics, and he rather goes out of his way to disown those aspects of genetics that are politically disrespectable. I feel that that's over and done with now, and we can forget about it and get on, and I feel he's still a little bit unnecessarily eager to distance himself from the bad aspects of the history of genetics. But I have a lot of time for him; I greatly respect him.

STEPHEN JAY GOULD: I like Steve Jones' work. I've read most of his sci-entific papers. I work on pulmonate snails, and he's one of the best in this little field. I don't know him very well. He's a very good scientist. He's followed the path of a media person, but in my professional world—snail biology—his science is very good.

Chapter 6

NILES ELDREDGE

"A Battle of Words"

DANIEL C. DENNETT: What Niles Eldredge wanted to show, and did show, along with Stephen Jay Gould, in their classic 1972 paper on punctuated equilibrium, was that the reigning assumption of their fellow paleontologists that the fossil records should show smooth gradual change over any timescale was wrong. It's very important that they pointed that out. What was even more important was that it didn't have the explanation that Darwin had given.

• • •

NILES ELDREDGE is a paleontologist; curator in the Department of Invertebrates at The American Museum of Natural History, in New York; author of Time Frames: The Rethinking of Darwinian Evolution and the Theory of Punctuated Equilibria *and* Unfinished Synthesis *(1985),* The Miner's Canary *(1991), and* Fossils *(1991), and* Reinventing Darwin *(1995).*

NILES ELDREDGE: Punctuated equilibria rests on a basic empirical claim, which is that once a species appears it typically doesn't change very much. If you're talking about marine invertebrates, that means five or ten million years. Yet evolution does, of course, occur, and the change seems to be associated with speciation events, in a formal sense. There's no intuitive reason why that should be true, because speciation is the setting up of new reproductive communities. It shouldn't have anything to do with adaptive change whatsoever, and yet it seems to.

I'm known for my work, in association with Steve Gould, on punctuated equilibria, also on fleshing out the hierarchical structure of biological systems: the ecological and genealogical twin hierarchies. It's basic ontology. Punctuated equilibria was a matter of putting Theodosius Dobzhansky's and Ernst Mayr's attention to the nature of species and discontinuities between them arising from the process of speciation, together with George Simpson's attention to evolutionary pattern in the fossil record. The resulting implications—one of which is the nature in general of large-scale biological systems—have occupied most of my attention.

There are several important paradoxes raised by the very notion of punctuated equilibria. One has to do with long-term evolutionary trends. If you look at long-term events in evolution within a group, like increased brain size in humans over four or five million years, the old model is that natural selection favors bigger brains, and so over four million years you get bigger and bigger brains. If you look at the fossil record, you do get bigger and bigger brains as you go through time, but it's a stepwise pattern, not a gradual thing. The raw statement of punctuated equilibria removed the old convenient element of directionality for long-term trends in the fossil record. But if you concede that there's a directionality over time, what is the explanation?

What we're saying is that species are entities. They have histories, they have origins, they have terminations, and they may or may not give rise to descendant species. They are individuals in the sense that human beings are individuals, albeit a very different kind of individuals. They're large-scale systems that have an element of reality to them, and that's a big departure in evolutionary biology. There are some adumbrations of it, but certainly it's not traditional. Our notion—sometimes called "species selection" or "species sorting" (a better term)—sees the differential origins and extinctions of species as an important additional element shaping the history of life, including the production of long-term evolutionary trends.

Species are real entities, spatiotemporally bounded, and they're information entities. Other kinds of entities do things. Ecological populations, for example, have niches; they function. Species don't function that way. They don't do things; they are, instead, information repositories. A species is not like an organism at all, but it's nonetheless a kind of entity that plays an important role in the evolutionary process.

This is an ontological shift. Geneticists remain underwhelmed with the notion that species are real entities, simply because they think that their data (generation-by-generation change within populations) don't demand it. So John Maynard Smith, whom I respect a great deal, won't respond to the gambit that species are real entities, with actual roles to play in the evolutionary process, simply because he doesn't personally find it interesting. Richard Dawkins is trying to force everything into a genic explanation. It's ships passing in the night. You don't need that kind of concept at all, if you're just dealing with generation by generation, running the natural-selection algorithm. I don't blame those who don't find this interesting. It's our job to make them understand that construing species in this way adds a valid and interesting element to evolutionary theory.

I call people like Steve and myself "the naturalists," in contrast with our gene-minded colleagues, the ultra-Darwinians. We try to capture the middle ground. The three main characters in the ultra-Darwinian camp would be Maynard Smith, George Williams, and

Richard Dawkins. All of us agree on the rudiments of evolutionary change: adaptive modification through natural selection. The one gloss on it that ultra-Darwinians have developed is that the fundamental dynamic underlying all of biotic nature is a competitive urge to leave copies of your genes behind. Making bits of information—genes—compete with one another makes evolutionary biology seem more like physics, so my accusing ultra-Darwinians of physics envy is probably the snottiest thing I could say about them.

Punctuated equilibria reasserts the importance of discontinuity in evolutionary discourse. Though it's usually the ultra-Darwinians who are cast in the role of *defensores fidei*, it was actually Mayr and Dobzhansky, as founders of the "modern synthesis," who originally managed to inject an element of discontinuity into the evolutionary discourse. So it's actually we naturalists who are defending a corner of orthodoxy here. As Dobzhansky said, Darwin established the validity of natural selection, and natural selection generates a spectrum of continuous variation. But nature is discontinuous. It's discontinuous (as Dobzhansky said in 1937) at the gene level and again at the species level. Most evolutionary biologists are population geneticists, and so they don't, as I just said, see the significance of intraspecific discontinuity. The data they handle aren't at the intraspecies level. They're not used to thinking about these problems, so a lot of that early discourse, hard-fought and hard-won, about the differences between species, defined as separate reproductive communities, doesn't enter into their universe. That's O.K., in a sense, because, what the hell, you talk about the stuff that impinges on your consciousness, and what your data are all about, and there's a lot to be said about the forces affecting the frequencies of genes on a generation-by-generation level within populations. But to restrict the discourse to species is to ignore other elements of biological organization, and ultra-Darwinians do so at their peril. Theirs is an incomplete description of biotic nature, rendering their theory simplistic and incomplete. It's disturbing that in his recent book *Natural Selection*, George Williams goes out of his way to stress that species are no special category of biological entity.

Ultra-Darwinians generally deny that they're genetic reductionists, but by anyone's definition they absolutely are. They try to explain the structure and history of large-scale systems purely in terms of relative gene frequencies. Social systems, economic systems, ecosystems, and so forth, all flow from this supposed competition among organisms—or even worse, among the genes. So what they're doing is actually playing fast and loose with a lot of work that's been done over the last fifty years establishing the actual nature of large-scale biological systems like species, like ecosystems—like social systems, for that matter—because they have a very gene-slanted view.

In a sense, I think it's intellectually incomplete, rather than dishonest. I always feel that if you're going to critique somebody you ought to know how to sing their song, and I feel like I spend a lot of time learning how to sing these guys' song. I don't see them turning around and learning the song that Steve and I and Elisabeth Vrba and Steven Stanley have been singing. I think they're so wrapped up in their own gene-centered world that they have an incomplete ontology of biological nature.

George Williams was the one who began taking evolution out of the passive mode and making it active. The translation of this is that organisms are out there competing, and although it looks like they're competing for food, they're competing for the opportunity to leave more genes behind. At the reproductive-biology level, it's a good description of nature. As a rubric to explain what's going on in biological nature in general, especially in large-scale biological systems, it falters—increasingly as you enter larger-scale systems and particularly as you address economic rather than reproductive phenomena.

Dawkins says in *The Selfish Gene* that ultimately we'll be able to understand the entire internal workings of ecosystems—the rules of assembly and what keeps them together—based on this particular principle of genic competition. It's an appeal to reductive thinking, and an attempt to turn into an active principle what Darwin was content to leave as a passive principle.

This is a subtle point. Ultra-Darwinians are reductionists, but only down to the genes-within-populations level. They're afraid of still lower levels: most population geneticists freak out when they hear a molecular biologist like Gabriel Dover talk about evolution! What we're saying is that there are more levels, both higher and lower, than in the traditional bailiwick of population genetics. Things are a little more complex, and we can specify to some degree what that complexity is.

If you read Dawkins' *The Blind Watchmaker*, it presents a seemingly adequate theory of why organisms appear to fit their environment so well—in other words, how natural selection shapes organismic adaptation. But on closer scrutiny, you find that there's absolutely nothing in there about why adaptive change occurs in evolution. There's really nothing about the context of adaptive change. It's just not addressed; it's not even an issue. It's just an in-principle argument. The algorithm is described in loving detail, and the supposition is that you just let the motor run and all the stuff we see—all that three-and-a-half-million-year history, those ten-million-odd species we have on Earth right now—simply falls out of that. The rest of it is mere detail. The important thing is to get the mechanism.

As I've said, we all basically agree on the statement of the mechanism, which Darwin established. Nobody's going to argue against it. Trying to make Steve Gould out as an anti-adaptationist is crap. Maynard Smith had a symposium in 1978, and as he himself says, the most visible thing that came out of it was the paper Steve wrote with Dick Lewontin called "The Spandrels of San Marco and the Panglossian Paradigm," saying that there's this adaptationist program and its proponents assume that every biological structure we look at was carefully chosen by natural selection. It's just an assumption, and hard to show rigorously, and there are an awful lot of other possibilities. That statement cast him in the role of being an anti-adaptationist.

The adaptationism of Maynard Smith, Williams, and Dawkins can be explained as follows: there's design in nature, organisms look like they're fairly well suited to the environment they find themselves in, and they're functioning pretty well out there. The only explanation for this state of affairs that makes any sense, other than the notion that a creator did it, is the process of evolution—particularly through natural selection, where the best-suited variants in the population tend, on average, to leave more copies of their genes behind than others less well endowed. Over a process of generations, nature will cull out and push things, giving the requisite variations. Those are the ground rules; everybody accepts that. What we naturalists are saying, in contrast, is that natural selection seems to produce adaptive change mainly in conjunction with true speciation—the sundering of an ancestral reproductive community ("species") into two or more descendant species.

A lot of the debate between Dawkins and Gould is trying to show each other who's brighter and more clever. I sometimes think they're almost deliberately misconstruing each other. It's a battle of words, a battle of wills to try to inform the literate public about who has the best approach to nature.

Steve has always been ready, willing, and able to jump in there and joust with the Dawkinses and Maynard Smiths of this world. It's taken me a longer time to get to that point. I focused first on reconciling patterns of stasis and change in the fossil record with the views of my immediate predecessors—Simpson the paleontologist, of course, but also Mayr the systematist, and even, oddly, Dobzhansky the consummate naturalist-turned-geneticist. I had to get past seeing Simpson, Mayr, and Dobzhansky in the Oedipal sense—as father figures whose work I had to correct. I had to come to understand that the punctuated-equilibria idea actually reconciles some serious discrepancies between Simpson, on the one hand, and Mayr and Dobzhansky on the other. Once done, the radical implications of punctuated equilibria stood out in bolder relief, and I could turn my attention more fully to the modern ultra-Darwinians. It's only now that I'm ready to take on people like Richard Dawkins and Maynard Smith and George Williams.

George Williams makes an interesting, if depressing, observation in *Natural Selection*. He says that a lot of the problems aren't so much solved as that people stop probing them and convince each other that they're solved well enough. We don't so much solve scientific problems as abandon them. That's kind of chilling, but it's a good description of what goes on. What I'm saying here is that there's no final resolution to the current debates. The millennial issues of human overpopulation, large-scale environmental destruction, and species loss probably will—probably should—distract us from the less pressing concerns of pure evolutionary theory. The issues will be picked up again someday, but by our intellectual descendants.

STEPHEN JAY GOULD: Niles Eldredge is my closest and dearest colleague in science. Whenever two people are so strongly connected, as we have been through punctuated equilibrium, people inevitably try to drive wedges, and yet although some attempts have been made no one has ever succeeded.

I receive more attacks, and people are often attacking other things that I stand for, because I have a more public reputation. But Niles and I don't agree on everything by any means. For example, he's a cladist and I'm not. We certainly disagree about a variety of technical aspects in hierarchical selection theory. He's very intrigued, for example, by a notion of parallel hierarchies, one genealogical and one which he calls economic, having to do with actual and overt competitions in nature. I myself, for a variety of technical reasons, think that the genealogical hierarchies, where causality resides, should be the ones we focus on. But Niles and I have worked together on this stuff for twenty-one years, and we're as close now as we've ever been. We were graduate students together. He's built one of the best Victorian cornet collections anywhere. He plays cornet and trumpet. He plays a mean jazz cornet.

DANIEL C. DENNETT: What Niles Eldredge wanted to show, and did show, along with Stephen Jay Gould, in their classic 1972 paper on punctuated equilibrium, was that the reigning assumption of their fellow paleontologists that the fossil records should show smooth gradual change over any timescale was wrong. It's very important that they pointed that out. What was even more important was that it didn't have the explanation that Darwin had given. Darwin had also been worried about the problem that the fossil record doesn't show lots of intermediate cases. He explained it in terms of imperfections of the fossil record: as if once we gather more data, unless we have bad luck, we'll eventually find all the intermediate cases and everything will be hunky-dory.

Eldredge and Gould showed, perhaps for the first time, that you should expect sudden transitions in even the most perfect fossil record. As Darwin himself had pointed out, most species—most lineages—should be in stasis most of the time anyway. Change doesn't happen continuously; it happens during brief periods, when there is a lot of change. These are also periods in which populations shift their locations, leaving their ancestral homes and invading new territory. And then a new equilibrium point is reached, after which you have another long period of stasis. Almost anywhere you look at the fossil record, which is both a spatial and a temporal cross-section, if you see any change at all, it should look like sudden arrivals.

But a lot of people have wanted to read something much more into it; they've wanted to say that the proof of punctuated equilibrium—let's grant that they've proved that there is punctuated equilibrium—is a refutation of Darwinian gradualism. It's nothing of the sort. Eldredge knows that, and doesn't deny it. Gould isn't so sure. Over the years, Gould has tried out various different ways in which what happens during the punctuation could be importantly non-Darwinian. I think I can show that each of these ways must be wrong. It's pretty clear that when you sort the issues out—and this isn't really a controversial opinion—there's no defensible revolutionary hypothesis about what's going on during punctuated equilibrium. But that's not the common perception of punctuated equilibrium among bystanders.

GEORGE C. WILLIAMS: Eldredge is a great paleontologist, and his recognition of stasis as an important conceptual challenge was a great advance. Stasis is the frequently seen stability of characters that we might think of as subject to rapid evolution. I'm not impressed with some of his recent work with Marjorie Grene, and I think he's made a conceptual muddle of things like sexual selection and levels of selection.

Gould and Eldredge, as the two fathers of punctuated equilibria, are of course often linked together. Gould is much more inclined to look—and perhaps more talented at looking—at big conceptual issues.

LYNN MARGULIS: Niles Eldredge, a broadly educated geologist and biologist, is a wonderful interpreter of the fossil record. His gorgeously illustrated book, *Fossils*, is a splendid introduction to the vertebrate and marine-animal fossil record. Furthermore, Eldredge recognizes how modern biology infringes and impinges upon everyday life, in "common myth," "commonsense knowledge," and other preconceived concepts that so many take for granted. What people in our culture take for granted is often diametrically opposed to what science, especially biology, tells us. Eldredge is one of the few people willing and able to interface between commonly held truths that are, in my opinion and in his opinion, gross misunderstandings and what science really tells us. He's a writer who truly communicates, and I want to see him keep writing.

W. DANIEL HILLIS: One of the first things I noticed about the evolution in the computer was that it didn't happen gradually. Nothing would happen for a long time, and then the world would reorganize itself and there would be a big change. Obviously that level of description connects to a group of biologists of which Eldredge is one of the leaders—the "Punc Eq" crowd. But when you look at it in detail you find that there are so many things that happen at so many different levels in biology that it's not clear that the phenomenon I was seeing corresponds exactly to the phenomenon he's describing. He's describing punctuated equilibrium on a grand scale of entire ecosystems, but it may be that that happens at a minor scale, too—even just within one evolving species. To me, punctuated equilibrium is totally noncontroversial; it happens at all kinds of different levels. It's one of those things that's completely obvious to anybody who's had a chance to play with a few million generations of evolution.

Chapter 7

LYNN MARGULIS

"Gaia Is a Tough Bitch"

RICHARD DAWKINS: *I greatly admire Lynn Margulis's sheer courage and stamina in sticking by the endosymbiosis theory, and carrying it through from being an unorthodoxy to an orthodoxy. I'm referring to the theory that the eukaryotic cell is a symbiotic union of primitive prokaryotic cells. This is one of the great achievements of twentieth-century evolutionary biology, and I greatly admire her for it.*

• • •

LYNN MARGULIS *is a biologist; Distinguished University Professor in the Department of Biology at the University of Massachusetts at Amherst; author of* The Origin of Eukaryotic Cells *(1970),* Early Life *(1981), and* Symbiosis in Cell Evolution *(2d ed., 1993). She is also the coauthor, with Karlene V. Schwartz, of* Five Kingdoms: An Illustrated Guide to the Phyla of Life on Earth *(2d ed., 1988) and with Dorion Sagan of* Microcosmos *(1986),* Origins Of Sex *(1986), and* Mystery Dance *(1991).*

LYNN MARGULIS: At any fine museum of natural history—say, in New York, Cleveland, or Paris—the visitor will find a hall of ancient life, a display of evolution that begins with the trilobite fossils and passes by giant nautiloids, dinosaurs, cave bears, and other extinct animals fascinating to children. Evolutionists have been preoccupied with the history of animal life in the last five hundred million years. But we now know that life itself evolved much earlier than that. The fossil record begins nearly four thousand million years ago! Until the 1960s, scientists ignored fossil evidence for the evolution of life, because it was uninterpretable.

I work in evolutionary biology, but with cells and microorganisms. Richard Dawkins, John Maynard Smith, George Williams, Richard Lewontin, Niles Eldredge, and Stephen Jay Gould all come out of the zoological tradition, which suggests to me that, in the words of our colleague Simon Robson, they deal with a data set some three billion years out of date. Eldredge and Gould and their many colleagues tend to codify an incredible ignorance of where the real action is in evolution, as they limit the domain of interest to animals—including, of course, people. All very interesting, but animals are very tardy on the evolutionary scene, and they give us little real insight into the major sources of evolution's creativity. It's as if you wrote a four-volume tome supposedly on world history but beginning in the year 1800 at Fort Dearborn and the founding of Chicago. You might be entirely correct about the nineteenth-century transformation of Fort Dearborn into a thriving lakeside metropolis, but it would hardly be world history.

By "codifying ignorance" I refer in part to the fact that they miss four out of the five kingdoms of life. Animals are only one of these kingdoms. They miss bacteria, protoctista, fungi, and plants. They take a small and interesting chapter in the book of evolution and extrapolate it into the entire encyclopedia of life. Skewed and limited in their perspective, they are not wrong so much as grossly uninformed.

Of what are they ignorant? Chemistry, primarily, because the language of evolutionary biology is the language of chemistry, and most of them ignore chemistry. I don't want to lump them all together, because, first of all, Gould and Eldredge have found out very clearly that gradual evolutionary changes through time, expected by Darwin to be documented in the fossil record, are not the way it happened. Fossil morphologies persist for long periods of time, and after stasis, discontinuities are observed. I don't think these observations are even debatable. John Maynard Smith, an engineer by training, knows much of his biology secondhand. He seldom deals with live organisms. He computes and he reads. I suspect that it's very hard for him to have insight into any group of organisms when he does not deal with them directly. Biologists, especially, need direct sensory communication with the live beings they study and about which they write.

Reconstructing evolutionary history through fossils—paleontology—is a valid approach, in my opinion, but paleontologists must work simultaneously with modern-counterpart organisms and with "neontologists"—that is, biologists. Gould, Eldredge, and Lewontin have made very valuable contributions. But the Dawkins-Williams-Maynard Smith tradition emerges from a history that I doubt they see in its Anglophone social context. Darwin claimed that populations of organisms change gradually through time as their members are weeded out, which is his basic idea of evolution through natural selection. Mendel, who developed the rules for genetic traits passing from one generation to another, made it very clear that while those traits reassort, they don't change over time. A white flower mated to a red flower has pink offspring, and if that pink flower is crossed with another pink flower the offspring that result are just as red or just as white or just as pink as the original parent or grandparent. Species of organisms, Mendel insisted, don't change through time. The mixture or blending that produced the pink is superficial. The genes are simply shuffled around to come out in different combinations, but those

same combinations generate exactly the same types. Mendel's observations are incontrovertible.

So J.B.S. Haldane, without a doubt a brilliant person, and R.A.
Fisher, a mathematician, generated an entire school of English-
speaking evolutionists, as they developed the neo-Darwinist popula-
tion-genetic analysis to reconcile two unreconcilable views: Darwin's
evolutionary view with Mendel's pragmatic, anti-evolutionary concept. They invented a language of population genetics in the 1920s to
1950s called neo-Darwinism, to rationalize these two fields. They
mathematized their work and began to believe in it, spreading the
word widely in Great Britain, the United States, and beyond. France
and other countries resisted neo-Darwinism, but some Japanese and
other investigators joined in the "explanation" activity.

Both Dawkins and Lewontin, who consider themselves far apart
from each other in many respects, belong to this tradition. Lewontin
visited an economics class at the University of Massachusetts a few
years ago to talk to the students. In a kind of neo-Darwinian jockey-
ing, he said that evolutionary changes are due to the Fisher-Haldane
mechanisms: mutation, emigration, immigration, and the like. At the
end of the hour, he said that none of the consequences of the details
of his analysis had been shown empirically. His elaborate cost-benefit
mathematical treatment was devoid of chemistry and biology. I asked
him why, if none of it could be shown experimentally or in the field,
he was so wedded to presenting a cost-benefit explanation derived
from phony human social-economic "theory." Why, when he himself
was pointing to serious flaws related to the fundamental assumptions,
did he want to teach this nonsense? His response was that there
were two reasons: the first was "P.E." "P.E.?," I asked. "What is
P.E.? Population explosion? Punctuated equilibrium? Physical education?" "No," he replied, "P.E. is 'physics envy,'" which is a syndrome
in which scientists in other disciplines yearn for the mathematically
explicit models of physics. His second reason was even more insidi-
ous: if he didn't couch his studies in the neo-Darwinist thought style
(archaic and totally inappropriate language, in my opinion), he
wouldn't be able to obtain grant money that was set up to support
this kind of work.

The neo-Darwinist population-genetics tradition is reminiscent of
phrenology, I think, and is a kind of science that can expect exactly

the same fate. It will look ridiculous in retrospect, because it is ridiculous. I've always felt that way, even as a more-than-adequate student of population genetics with a superb teacher—James F. Crow, at the University of Wisconsin, Madison. At the very end of the semester, the last week was spent on discussing the actual observational and experimental studies related to the models, but none of the outcomes of the experiments matched the theory.

I've been critical of mathematical neo-Darwinism for years; it never made much sense to me. We were all told that random mutations—most of which are known to be deleterious—are the main cause of evolutionary change. I remember waking up one day with an epiphanous revelation: I am not a neo-Darwinist! It recalled an earlier experience, when I realized that I wasn't a humanistic Jew.

Although I greatly admire Darwin's contributions and agree with most of his theoretical analysis and I am a *Darwinist*, I am not a *neo-Darwinist*. One of Darwin's major insights is the recognition that all organisms are related by common ancestry. Today direct evidence for common ancestry—genetic, chemical, and otherwise—is overwhelming. Populations of organisms grow and reproduce at rates that are not sustainable in the real world, and therefore many more die or fail to reproduce than actually complete their life histories. The fact that all the organisms that are born or hatched or budded off do not and cannot possibly survive *is* natural selection. Observable inherited variation appears in all organisms that are hatched, born, budded off, or produced by division, and some variants do outgrow and outreproduce others. These are the tenets of Darwinian evolution and natural selection. All thinking scientists are in complete agreement with these basic ideas, since they're supported by vast amounts of evidence.

Neo-Darwinism is an attempt to reconcile Mendelian genetics, which says that organisms do not change with time, with Darwinism, which claims they do. It's a rationalization that fuses two somewhat flawed traditions in a mathematical way, and that is the beginning of the end. Neo-Darwinist formality uses an arithmetic and an algebra that is inappropriate for biology. The language of life is not ordinary arithmetic and algebra; the language of life is chemistry. The practicing neo-Darwinists lack relevant knowledge in, for example, microbiology, cell biology, biochemistry, molecular biology, and cytoplasmic genetics. They avoid biochemical cytology and microbial ecology. This is comparable to attempting a critical analysis of Shakespeare's Elizabethan phraseology and idiomatic expression in Chinese, while ignoring the relevance of the English language!

The neo-Darwinists say that variation originates from random mutation, defining mutation as any genetic change. By randomness they mean that characters appear randomly in offspring with respect to selection: if an animal needs a tail, it doesn't develop this tail because it needs it; rather, the animal randomly develops all sorts of changes and those with tails survive to produce more offspring. H.J. Muller, in the 1920s, discovered that not only do X rays increase the fruit-fly mutation rate, but even if fruit flies are isolated completely from X rays, solar radiation, and other environmental perturbation, a spontaneous mutation rate can be measured. Inherited variants do appear spontaneously; they have nothing to do with whether or not they're good for the organism in which they appear. Mutation was then touted as the source of variation—that upon which natural selection acted—and the neo-Darwinian theory was declared complete. The science remaining required filling in the gaps in a "theory" with very few holes.

From many experiments, it is known that if mutagens like X rays or certain chemicals are presented to fruit flies, sick and dead flies result. No new species of fly appears—that is the real rub. Everyone agrees that such mutagens produce inherited variation. Everyone agrees that natural selection acts on this variation. The question is, From where comes the useful variation upon which selection acts? This problem has not yet been solved. But I claim that most significant inherited variation comes from mergers—from what the Russians, especially Konstantin S. Mereschkovsky, called *symbiogenesis* and the American Ivan Emanuel Wallin called *symbionticism*. Wallin meant by the term the incorporation of microbial genetic systems into progenitors of animal or plant cells. The new genetic system—a merger between microbe and animal cell or microbe and plant cell— is really different from the ancestral cell that lacks the microbe. Analogous to improvements in computer technology, instead of starting from scratch to make all new modules again, the symbiosis idea is an interfacing of preexisting modules. Mergers result in the emergence of new and more complex beings. I doubt new species form just from random mutation.

Symbiosis is a physical association between organisms, the living together of organisms of different species in the same place at the same time. My work in symbiosis comes out of cytoplasmic genetic systems. We were all taught that the genes were in the nucleus and that the nucleus is the central control of the cell. Early in my study of genetics, I became aware that other genetic systems with different inheritance patterns exist. From the beginning, I was curious about these unruly genes that weren't in the nucleus. The most famous of them was a cytoplasmic gene called "killer," which, in the protist *Paramecium aurelia*, followed certain rules of inheritance. The killer gene, after twenty years of intense work and shifting paradigmatic ideas, turns out to be in a virus inside a symbiotic bacterium. Nearly all extranuclear genes are derived from bacteria or other sorts of microbes. In the search for what genes outside the nucleus really are, I became more and more aware that they're cohabiting entities, live beings. Live small cells reside inside the larger cells. Understanding that led me and others to study modern symbioses.

Symbiosis has nothing to do with cost or benefit. The benefit/cost people have perverted the science with invidious economic analogies. The contention is not over modern symbioses, simply the living together of unlike organisms, but over whether "symbiogenesis"— long-term symbioses that lead to new forms of life—has occurred and is still occurring. The importance of symbiogenesis as a major source of evolutionary change is what is debated. I contend that symbiogenesis is the result of long-term living together—staying together, especially involving microbes—and that it's the major evolutionary innovator in all lineages of larger nonbacterial organisms.

In 1966, I wrote a paper on symbiogenesis called "The Origin of Mitosing [Eukaryotic] Cells," dealing with the origin of all cells except bacteria. (The origin of bacterial cells is the origin of life itself.) The paper was rejected by about fifteen scientific journals, because it was flawed; also, it was too new and nobody could evaluate it. Finally, James F. Danielli, the editor of *The Journal of Theoretical Biology*, accepted it and encouraged me. At the time, I was an absolute nobody, and, what was unheard of, this paper received eight hundred reprint requests. Later, at Boston University, it won an award for the year's best faculty publication. I was only an instructor at the time, so my Biology Department colleagues reacted to the commotion and threw a party. But it was more of "Isn't this cute," or "It's so abstruse

that I don't understand it, but others think it worthy of attention."
Even today most scientists still don't take symbiosis seriously as an
evolutionary mechanism. If they were to take symbiogenesis seriously, they'd have to change their behavior. The only way behavior
changes in science is that certain people die and differently behaving
people take their places.

Next, expanding the journal article, after ten years of research
and six weeks of intense writing, I produced a book called *The Origin
of Eukaryotic Cells.* Even under contract, it was rejected by Academic Press. Finally, in 1970, the revised and improved work was
published by Yale University Press. Now called *Symbiosis in Cell
Evolution*, the most recent version of the statement is in a second—
really a third—edition. Published by W.H. Freeman in 1993, that
book is my life's work. It details the role of symbiosis in the evolution
of cells, which leads directly to the origin of mitotic cell division and
meiotic sexuality. My major thrust is how different bacteria form
consortia that, under ecological pressures, associate and undergo
metabolic and genetic change such that their tightly integrated communities result in individuality at a more complex level of organization. The case in point is the origin of nucleated (protoctist, animal,
fungal, and plant) cells from bacteria.

While Gould and the others tend to believe that species only di-
verge from one another, I claim that—more important in generation
of variation—species form new composite entities by fusion and
merger. Symbiogenesis is an extremely important mechanism of evolution. Symbiogenesis analysis impacts on developmental biology, on
taxonomy and systematics, and on cell biology; it hits some thirty
subfields of biology, and even geology. Symbiogenesis has many implications, which is part of the reason it is controversial. Most people
don't like to hear that what they have been doing all these years is
barking up the wrong tree.

My argument is radical only to the extent that it inspires scientists to change their status quo about many issues. To take seriously
our *Five Kingdoms* concept (the book by Karlene V. Schwartz and
me is based on work by Robert H. Whittaker and Herbert F.
Copeland) a school or a publisher would have to change its catalog. A
supplier has to relabel all its drawers and cabinets. Departments
must reorganize their budget items, and NASA, the National Science

Foundation, and various museums have to change staff titles and program-planning committees. The change from "plants versus animals" to the five kingdoms (bacteria, protoctista, animals, fungi, and plants) has such a profound implication for every aspect of biology as a social activity that resistance to accept it abounds. Scientists and those who pay them have to dismiss or ignore this potential reorganization because accepting the shifting boundaries and new alliances is strange and costly. It is far easier to stay with obsolete intellectual categories.

For more than a billion years, the only life on this planet consisted of bacterial cells, which, lacking nuclei, are called prokaryotes, or prokaryotic cells. They looked very much alike, and from the human-centered vantage point seem boring. However, bacteria are the source of reproduction, photosynthesis, movement—indeed, all interesting features of life except perhaps speech! They're still with us in large diversity and numbers. They still rule Earth. At some point, a new more complex kind of cell appeared on the scene, the eukaryotic cell, of which plant and animal bodies are composed. These cells contain certain organelles, including nuclei. Eukaryotic cells with an individuated nucleus are the building blocks of all familiar large forms of life. How did that evolution revolution occur? How did the eukaryotic cell appear? Probably it was an invasion of predators, at the outset. It may have started when one sort of squirming bacterium invaded another—seeking food, of course. But certain invasions evolved into truces; associations once ferocious became benign. When swimming bacterial would-be invaders took up residence inside their sluggish hosts, this joining of forces created a new whole that was, in effect, far greater than the sum of its parts: faster swimmers capable of moving large numbers of genes evolved. Some of these newcomers were uniquely competent in the evolutionary struggle. Further bacterial associations were added on, as the modern cell evolved.

One kind of evidence in favor of symbiogenesis in cell origins is mitochondria, the organelles inside most eukaryotic cells, which have their own separate DNA. In addition to the nuclear DNA, which is the human genome, each of us also has mitochondrial DNA. Our mitochondria, a completely different lineage, are inherited only from our mothers. None of our mitochondrial DNA comes from our fathers. Thus, in every fungus, animal, or plant (and in most protoctists), at least two distinct genealogies exist side by side. That, in itself, is a clue that at some point these organelles were distinct microorganisms that joined forces.

David Luck and John Hall, research geneticists at Rockefeller University, recently made an astounding discovery that I more or less predicted twenty-five years ago. They demonstrated by well-developed techniques something they were not even seeking: a peculiar DNA—outside the nucleus of the cell, outside the chloroplast, and outside the mitochondria. This extranuclear DNA, these genes outside the nucleus, can be interpreted as remnants of ancient, invasive, squirming bacteria whose aggressive association presaged the merger.

If their discovery is correct—and at least three teams of researchers have disputed it—then the nonnuclear genetic system Hall and Luck revealed in green algae may represent the stripped-down remnants of bacteria inside all of us. The growth, reproduction, and communication of these moving, alliance-forming bacteria become isomorphic with our thought, with our happiness, our sensitivities and stimulations. If mine is a correct view, it organizes a great deal of knowledge. There are unambiguous ways of testing the main points. The implication is that we are literally inhabited by highly motile remnants of an ancient bacterial type that have become, in every sense, a part of ourselves. These thriving partial beings represent the physical basis of *anima*: soul, life, locomotion; an advocation of materialism in the crassest sense of the word. Put it this way: a purified chemical is prepared from brain and added to another purified chemical. These two chemicals—two different kinds of motile proteins— together crawl away, they locomote. They move all by themselves. Biochemists and cell biologists can show us the minimal common denominator of movement, locomotion. *Anima*. Soul. These moving proteins I interpret as the remains of the swimming bacteria incorporated by beings who became our ancestors as they became us.

The minimal-movement system is so physically and chemically characterizable that complete consensus exists that "motility proteins" are composed of typical carbon-hydrogen bonds, and so forth. All the details are agreed upon by cell biologists and biochemists. But I think an understanding of the extent to which the evolutionary origin involved symbiogenesis must be acknowledged. Such acknowledgment will lead to new awareness of the physical basis of thought. Thought and behavior in people are rendered far less mysterious when we realize that choice and sensitivity are already exquisitely developed in the microbial cells that became our ancestors. Even philosophers will be inspired to learn about motility proteins. Scientists and nonscientists will be motivated to learn enough chemistry, microbiology, evolutionary biology, and paleontology to understand the relevance of these fields to the deep questions they pose.

My primary work has always been in cell evolution, yet for a long time I've been associated with James Lovelock and his Gaia hypothesis. In the early seventies, I was trying to align bacteria by their metabolic pathways. I noticed that all kinds of bacteria produced gases. Oxygen, hydrogen sulfide, carbon dioxide, nitrogen, ammonia—more than thirty different gases are given off by the bacteria whose evolutionary history I was keen to reconstruct. Why did every scientist I asked believe that atmospheric oxygen was a biological product but the other atmospheric gases—nitrogen, methane, sulfur, and so on—were not? "Go talk to Lovelock," at least four different scientists suggested. Lovelock believed that the gases in the atmosphere were biological. He had, by this time, a very good idea of which live organisms were probably "breathing out" the gases in question. These gases were far too abundant in the atmosphere to be formed by chemical and physical processes alone. He argued that the atmosphere was a physiological and not just a chemical system.

The Gaia hypothesis states that the temperature of the planet, the oxidation state and other chemistry of all of the gases of the lower atmosphere (except helium, argon, and other nonreactive ones) are produced and maintained by the sum of life. We explored how this could be. How could the temperature of the planet be regulated by living beings? How could the atmospheric gas composition—the 20-percent oxygen and the one to two parts per million methane, for example—be actively maintained by living matter?

It took me days of conversation even to begin to understand Lovelock's thinking. My first response, just like that of the neo-Darwinists, was "business as usual." I would say, "Oh, you mean that organisms adapt to their environment." He would respond, very sweetly, "No, I don't mean that." Lovelock kept telling me what he really meant, and it was hard for me to listen. Since his was a new idea, he hadn't yet developed an appropriate vocabulary. Perhaps I helped him work out his explanations, but I did very little else.

The Gaia hypothesis is a biological idea, but it's not human-centered. Those who want Gaia to be an Earth goddess for a cuddly, furry human environment find no solace in it. They tend to be critical or to misunderstand. They can buy into the theory only by misinterpreting it. Some critics are worried that the Gaia hypothesis says the environment will respond to any insults done to it and the natural systems will take care of the problems. This, they maintain, gives industries a license to pollute. Yes, Gaia will take care of itself; yes, environmental excesses will be ameliorated, but it's likely that such restoration of the environment will occur in a world devoid of people.

Lovelock would say that Earth is an organism. I disagree with this phraseology. No organism eats its own waste. I prefer to say that Earth is an ecosystem, one continuous enormous ecosystem composed of many component ecosystems. Lovelock's position is to let the people believe that Earth is an organism, because if they think it is just a pile of rocks they kick it, ignore it, and mistreat it. If they think Earth is an organism, they'll tend to treat it with respect. To me, this is a helpful cop-out, not science. Yet I do agree with Lovelock when he claims that most of the things scientists do are not science either. And I realize that by taking the stance he does he is more effective than I am in communicating Gaian ideas.

If science doesn't fit in with the cultural milieu, people dismiss science, they never reject their cultural milieu! If we are involved in science of which some aspects are not commensurate with the cultural milieu, then we are told that our science is flawed. I suspect that all people have cultural concepts into which science must fit. Although I try to recognize these biases in myself, I'm sure I cannot entirely avoid them. I try to focus on the direct observational aspects of science.

Gaia is a tough bitch—a system that has worked for over three billion years without people. This planet's surface and its atmosphere and environment will continue to evolve long after people and prejudice are gone.

DANIEL C. DENNETT: One of the most beautiful ideas I've ever encountered is Lynn Margulis's idea about the birth of the eukaryotic cell through a transformation of what started out as a parasitic infestation of one cell by another. When she first proposed it, she was scoffed at, laughed at, and it's delicious that this is now pretty well accepted as a major, major theoretical development. I think of her as one of the heroes of twentieth-century biology.

Some of her recent popular writing disturbs me, because I think she's trying to take that wonderful idea and harness it as a political idea, stressing cooperation over competition. But that seems to me to be a mistake. Yes, the eukaryotic revolution was an instance in which what began as competition evolved into what is fundamentally a cooperative arrangement. That's its beauty, but precisely what it doesn't show is that cooperation is the norm or that cooperation is always good or that it's always possible. It's the rare and wonderful thing that enabled multicellular life to take off. But you can't read into it any message such as that nature is fundamentally cooperation; it isn't.

GEORGE C. WILLIAMS: I'm probably being unfair, but I would say that Lynn Margulis is very much afflicted with a kind of "God-is-good" syndrome, in that she wants to look out there at nature and see something benign and benevolent and ultimately wholesome and worth having. Whereas I look out there with Tennyson and see things red in tooth and claw. In other words, it's a bloody mess out there.

She likes to look out there and see cooperation and things being nice to each other. This culminates in this Gaia idea. There's this entity—we will not make it a god or a goddess, let's say that the implication is just a metaphor. But that's what she wants to see, and therefore, come what may, that's what she's going to see. She could say the same about me—that I think "God is evil," and I look out there at His creations and see nothing but evil. Time will tell, and will show that my approach is more fruitful in generating predictions about discoveries we're going to make.

LEE SMOLIN: Lynn Margulis has been for many years one of my scientific heroes. She is, in my opinion, one of the great living American scientists. She has this ability, which I think of as the characteristic of the best scientists from the American tradition, of thinking in broadly significant and original ways while staying very close to nature—to the ground, so to speak. Richard Feynman was another like this. You couldn't speak to him about physics without speaking about nature. I can't judge her effect on biologists, except to say that I imagine most of them haven't yet caught up with her, but I can certainly say, as a physicist, that she's had a dramatic impact on how I think about biology. Three aspects of her vision—the importance of symbiosis in evolution, the Gaia hypothesis, and the view that the whole living world is an elaboration of microbial life—are, I believe, extremely important for understanding the relationship of the living world to the physical world at large.

I admired her for many years from her writings and from hearsay. I was very fortunate to meet her two years ago. At a dinner party, I witnessed her defend the Gaia hypothesis against what another biologist present had said in print. She had the unfortunate person cornered; she was able to quote, word for word from memory, what he'd said, and she was very intent on having him see why it was wrong. I must say that when I witnessed this conversation I was reminded of the accounts written of Galileo when he came to Rome, in which he is described as defending the Copernican hypothesis at dinner parties in the houses of the great families there. I saw in her the same confidence in her vision, together with impatience at those who can't think as openly or as broadly but instead choose to misunderstand the new ideas. I've thought for many years that we as yet barely understand the implications of Darwin's discovery that we evolved via natural selection. I'm sure that Lynn Margulis has seen further than most what this means for our view of the natural world and our relationship to it.

One thing I can't understand is the animosity among the different evolutionary theorists, such as Lynn Margulis, Richard Dawkins, and others. The idea that the world has evolved by variation and selection is, as far as I can tell, completely consistent with both the idea that symbiosis is a major mechanism of evolution and the idea that the whole biosphere functions as a single organism with mechanisms of self-regulation of climate and various cycles. It seems to me that rather than being contradictory, both aspects must be necessary. The living world must be a single self-organized entity to have come to exist at all, and the only way such complexity and astounding novelty can arise is by random variation and natural selection. As a physicist, I feel that the little we've so far been able to observe about how self-organized systems work points to the necessity of both aspects. For example, biologists seem to be endlessly arguing about the scale on which natural selection operates. Does it operate on the ecosystem, on the species, on the individual, on the gene? The one key lesson about self-organized systems that physicists have learned is that they're what we call critical systems, which are systems in which significant correlations are evolving on every possible scale. So the answer, I imagine, is that evolution must be taking place simultaneously on a large varieties of scales. Of course, the information is coded on one scale—that of the gene. But it's expressed over every scale from the individual cell to the biosphere as a whole. Thus, the probability that a gene will be reproduced is influenced by its effect on every scale, which means that evolution can act at every scale.

Of course, it would be good to have a general theory of self-organized systems which could serve as the starting point for such a discussion—and, indeed, for developing a general understanding of the biological world and its evolution. This has been a dream of physicists for a long time. It's certainly a dream of mine and of many other people. I think that just recently we can say we're beginning to uncover some concepts, such as self-organized criticality, which could play a role in such a theory.

MARVIN MINSKY: The animals we know today didn't evolve from scratch but, with the exception of bacteria, virtually all of them are fusions of three or four more primitive animals. That's how we got the way we are.

NILES ELDREDGE: Lynn is marvelous. I hope I'm not being too Pollyanna-ish, but her notion of the symbiotic origin of the eukaryotic cell was probably the grandest idea in modern biology. Lynn was put down as having had a really crazy idea, and, of course, we can relate to that. Now it's taught in all the textbooks as the self-evident truth. It was a marvelous thing.

Her involvement with Gaia has been more messy. This is James Lovelock's notion that the earth is a living organism—at least that's the strong form of it. Her commentaries on evolutionary biology sometimes miss the mark. She, like me and a lot of other people, thinks that the metaphor of competition for reproductive success is overdone in the ultra-Darwinian paradigm; but, on the other hand, there's no question that there's competition in nature, and she's trying to stress cooperation.

RICHARD DAWKINS: I greatly admire Lynn Margulis's sheer courage and stamina in sticking by the endosymbiosis theory, and carrying it through from being an unorthodoxy to an orthodoxy. I'm referring to the theory that the eukaryotic cell is a symbiotic union of primitive prokaryotic cells. This is one of the great achievements of twentieth-century evolutionary biology, and I greatly admire her for it.

I first met Lynn some years ago at a conference in the South of France, and I think we got on rather well together. I have since, when I've met her, found her extremely obstinate in argument. I have the feeling that she's the kind of person who just knows she's right and doesn't listen to argument. Whereas I think I actually do listen—and perhaps change my mind if someone presents a convincing argument—I get the feeling that she does not. That may be unfair, and in the case of the theory of the origin of the eukaryotic cell, she was right to be obstinate. She's turned out, probably, to be right, but that doesn't mean she's always right. And I suspect that she isn't always right.

The Gaia hypothesis is a good example of that. I don't think Lovelock was clear—in his first book, at least—on the kind of natural-selection process that was supposed to put together the adaptive unit, which in his case was the whole world. If you're going to talk about a unit at any level in the hierarchy of life as being adaptive, then there has to be some sort of selection going on among self-replicating information. And we have to ask, What is the equivalent of DNA? What are the units of code? What are the units of copy-me code which are being replicated?

I don't think for a moment that it occurred to Lovelock to ask himself that question. And so I'm skeptical of the rhetoric of the Gaia hypothesis, when it comes down to particular applications of it, like explaining the amount of methane there is in the atmosphere, or saying there will be some gas produced by bacteria which is good for the world at large and so the bacteria go to the trouble of producing it, for the good of the world. That can't happen in a Darwinian world, as long as we think that natural selection is going on at the level of individual bacterial genes. Because those individual bacteria who don't put themselves to the trouble of manufacturing this gas for the good of the world will do better. Of course, if the individual bacteria who manufacture the gas are really doing themselves better by doing so, and the gas is just an incidental consequence, obviously I have no problem with that, but in that case you don't need a Gaia hypothesis to explain it. You explain it at the level of what's good for the individual bacteria and their genes.

FRANCISCO VARELA: I consider Lynn Margulis one of the brightest and most important biologists since the geneticists of the 1920s, when Thomas H. Morgan and J.B.S. Haldane contributed to setting the basis of evolutionary biology on cellular grounds.

Since then, the single interesting piece that puts biology together at all levels—from geology to cell biology to molecular biology to evolution—is Lynn. One of the key ideas she introduced in the 1970s was evolution by symbiosis. Nobody believed her. By now that idea is pretty well established, and it's opened up a new way of thinking about the relationships we see in microorganisms. It also helps us to understand the importance of the microworld. She's very original. Her book *Symbiosis in Cell Evolution* is one of the classics of biology in the twentieth century.

I do have some criticisms. In recent years, she has taken some of her important scientific ideas into a more cultural sphere, towards a human cultural interpretation. This is bad. Her story on the origins of sex, in *Mystery Dance*, written with Dorion Sagan, for example, is naive, full of clichés, and devoid of historical perspective. I read Richard Dawkins' famous attack in his review in *Nature,* and I must say that for once I agree entirely with Dawkins. It's unfortunate that she has veered into some weird second stage.

One of the reasons I got to know her is that she was one of the first biologists who appreciated the work I did with Humberto Maturana back in the seventies, on understanding the basic biological cellular organization of an autopoietic system. She immediately took it up and incorporated it into her work. That was very important to me, because it meant that I was not, as we say in Spanish, completely outside the pot.

W. DANIEL HILLIS: Most of the science that gets done gets done within a rigid set of rules, where you know exactly who your peers are, and things get evaluated according to a very strict set of standards. That works, when you're not trying to change the structure. It works in what Stephen Gould calls incremental science. But when you try to change the structure, that system doesn't work very well. When you try to do something that doesn't fit into a discipline or a standard theory, you usually make some enemies. Lynn Margulis is an example of somebody who didn't follow the rules and pissed a lot of people off.

She had a way of looking at symbiosis which didn't fit into the popular theories and structure. In the minds of many people, she went around the powers that be and took her theories directly to the public, which annoyed them all. It particularly annoyed them because she turned out to be right. If it's a sin to take your theories to the public, then it is a double sin to take your theories to the public and be right.

Part Two

A COLLECTION

OF KLUDGES

One of the central metaphors of the third culture is computation. The computer does computation and the mind does computation. To understand what makes birds fly, you may look at airplanes, because there are principles of flight and aerodynamics that apply to anything that flies. That is how the idea of computation figures into the new ways in which scientists are thinking about complicated systems.

At first, people who wanted to be scientific about the mind tried to treat it by looking for fundamentals, as in physics. We had waves of so-called mathematical psychology, and before that psychologists were trying to find a simple building block—an "atom"—with which to reconstruct the mind. That approach did not work. It turns out that minds, which are brains, are extremely complicated artifacts of natural selection, and as such they have many emergent properties that can best be understood from an engineering point of view.

We are also discovering that the world itself is very "kludgey"; it is made up of curious Rube Goldberg mechanisms that do cute tricks. This does not sit well with those who want science to be crystalline and precise, like Newton's pure mathematics. The idea that nature might be composed of Rube Goldberg machines is deeply offensive to people who have a strong esthetic drive—those who say that science must be beautiful, that it must be pure, that everything should be symmetrical and deducible from first principles. That esthetic has been a great motivating force in science, since Plato.

Counteracting it is the esthetic that emerges from this book—the esthetic that says the beauties of nature come from the interaction of mind-boggling complexities, and that it is complexity essentially most of the way down. The computational perspective—machines made out of machines made out of machines—is on the ascendant. There is a lot of talk about machines in this book.

Marvin Minsky is the leading light of AI—that is, artificial intelligence. He sees the brain as a myriad of structures. Scientists who, like Minsky, take the strong AI view believe that a computer model of the brain will be able to explain what we know of the brain's cognitive abilities. Minsky identifies consciousness with high-level, abstract thought, and believes that in principle machines can do everything a conscious human being can do.

Roger Schank is a computer scientist and cognitive psychologist who has worked in the AI field for twenty years. Like Minsky, he takes the strong AI view, but rather than trying to build an intelligent machine he wants to deconstruct the human mind. He wants to know, in particular, how natural language—one's mother tongue—is processed, how memory works, and how learning occurs. Schank thinks of the human mind as a learning device, and he thinks that it is being taught in the wrong way. He is something of a gadfly; he deplores the curriculum-based, drill-oriented methods in today's schools, and his most recent contributions have been in the area of education, looking at ways to use computers to enhance the learning process.

The philosopher Daniel C. Dennett is interested in consciousness, and his view of it, similar to Minsky's, is as high-level, abstract thinking. He is known as the leading proponent of the computational model of the mind; he has clashed with philosophers such as John Searle who maintain that the most important aspects of consciousness—intentionality and subjective quality—can never be computed. He is the philosopher of choice of the AI community. In his more recent work, he has turned to what he calls "Darwin's dangerous idea"; he is squarely in the ultra-Darwinist camp of George C. Williams and Richard Dawkins, and he has with great energy mustered a serious critique of the scientific ideas of Stephen Jay Gould.

Nicholas Humphrey, a research psychologist who twenty years ago breathed life into the newly developing field of evolutionary psychology with his theory about "the social function of intellect," here discusses his more recent ideas on the nature of phenomenal consciousness. Unlike Dennett, who sees the role of philosophers as disabusing people of their "primitive" ideas about the nature of consciousness, Humphrey believes that we should take these primitive intuitions at face value. If people say that the problem is what it "feels like" to be conscious, then the problem is indeed to explain "feeling." Humphrey and Dennett are a pair of bookends. Some regard Humphrey as a "romantic scientist," more interested in storytelling than in getting at the scientific facts. But he would probably not agree that there is a hard and fast line between facts and stories.

Francisco Varela, an experimental and theoretical biologist, studies what he terms "emergent selves" or "virtual identities." His is an immanent view of reality, based on metaphors derived from self-organization and Buddhist-inspired epistemology rather than on those derived from engineering and information science. He presents a challenge to the traditional AI view that the world exists independently of the organism, whose task is to make an accurate model of that world—to "consult" before acting. Varela's nonrepresentationalist world—or perhaps "world-as-experienced"—has no independent existence but is itself a product of interactions between organisms and environment. He first became known for his theory of autopoiesis ("self-production"), which is concerned with the active self-maintenance of living systems whose identities remain constant while their components continually change. Varela is tough to categorize. He is a neuroscientist who has become an immunologist. He is well informed about cognitive science and is a radical critic of it, because he is a believer in "emergence"—not the vitalist idea promulgated in the 1920s (that of a magical property that emerges inexplicably from lower mechanical operations) but the idea that the whole appears as a result of the dynamics of its component parts. He thinks that classic computationalist cognitive science is too simple-mindedly mechanistic. He is knowledgeable and romantic at the same time.

The experimental psychologist Steven Pinker is a unifier, some-one who ties a lot of big ideas together—from evolutionary theory to consciousness to the language "instinct." He has studied visual cogni-tion and language acquisition in the laboratory, and was one of the first to develop computational models of how children learn the words and grammar of their first language. He has merged Chom-skyan ideas about the innate character of language with Darwinian explanations, such as adaptation and natural selection. Pinker wrote one of the most influential critiques of neural-network models of the mind. He takes the position that even the simplest tasks that hu-mans perform—picking up a pencil, responding to a color, recogniz-ing a friend's face—are extraordinary engineering feats and beyond the capabilities of any current software designers. He believes that the brain has to have a set of specialized tools and that there is no one general-purpose learning machine capable of duplicating its feats.

The mathematical physicist Roger Penrose attempts to link the quantum and the classical world in physics. He believes that a lot of what the brain does is what you could do on a computer, but he posits that the actions of consciousness are something different. To allow for the brain's capacity to know mathematical truth, he thinks, physics must include a noncomputable element, which he surmises will emerge in a theory of quantum gravity. This leads many of his colleagues to wonder if he is a physicist gone astray. Indeed, his in-ternational best-seller, *The Emperor's New Mind*, is considered by some of them to be *The Emperor's New Book*. Yet the reading public had a very different reaction—the book reached the #7 position on the *New York Times'* best-seller list. Why? It is a tour de force through the realm of physics and is filled with pages of mathematical formulas and equations. In claiming that the human mind is not equatable to a machine, Penrose obviously said what many people wanted to hear: they may well have bought the book as a talisman in support of that which they would have be true. Among his scientific colleagues there is less backing for what many consider to be a radical theory. Yet Penrose retains their respect, because of his utter honesty and inquiring mind. In *Shadows of the Mind*, his sequel, he addresses the scientific arguments concerning his theories of mind in a rigorous fashion.

Chapter 8

MARVIN MINSKY

"Smart Machines"

ROGER SCHANK: *Marvin Minsky is the smartest person I've ever known. He's absolutely full of ideas, and he hasn't gotten one step slower or one step dumber. One of the things about Marvin that's really fantastic is that he never got too old. He's wonderfully childlike. I think that's a major factor explaining why he's such a good thinker. There are aspects of him I'd like to pattern myself after. Because what happens to some scientists is that they get full of their power and importance, and they lose track of how to think brilliant thoughts. That's never happened to Marvin.*

•   •   •

MARVIN MINSKY *is a mathematician and computer scientist; Toshiba Professor of Media Arts and Sciences at the Massachusetts Institute of Technology; cofounder of MIT's Artificial Intelligence Laboratory, Logo Computer Systems, Inc., and Thinking Machines, Inc.; laureate of the Japan Prize (1990), that nation's highest distinction in science and technology; author of eight books, including* The Society of Mind *(1986).*

MARVIN MINSKY: Like everyone else, I think most of the time. But mostly I think about thinking. How do people recognize things? How do we make our decisions? How do we get our new ideas? How do we learn from experience? Of course, I don't think only about psychology. I like solving problems in other fields—engineering, mathematics, physics, and biology. But whenever a problem seems too hard, I start wondering why that problem seems so hard, and we're back again to psychology! Of course, we all use familiar self-help techniques, such as asking, "Am I representing the problem in an unsuitable way," or "Am I trying to use an unsuitable method?" However, another way is to ask, "How would I make a machine to solve that kind of problem?"

A century ago, there would have been no way even to start thinking about making smart machines. Today, though, there are lots of good ideas about this. The trouble is, almost no one has thought enough about how to put all those ideas together. That's what I think about most of the time.

The technical field of research toward machine intelligence really started with the emergence in the 1940s of what was first called cybernetics. Soon this became a main concern of several different scientific fields, including computer science, neuropsychology, computational linguistics, control theory, cognitive psychology, artificial intelligence—and, more recently, the new fields called connectionism, virtual reality, intelligent agents, and artificial life.

Why are so many people now concerned with making machines that think and learn? It's clear that this is useful to do, because we already have so many machines that solve so many important and interesting problems. But I think we're motivated also by a negative reason: the sense that our traditional concepts about psychology are no longer serving us well enough. Psychology developed rapidly in the early years of this century, and produced many good theories about the periphery of psychology—notably, about certain aspects of perception, learning, and language. But experimental psychology never told us enough about issues of more central concern—about thinking, meaning, consciousness, or feeling.

The early history of modern psychology also produced some higher-level conceptions that promised—at least in principle—to explain a great deal more. These included, for example, the kinds of theories proposed by Freud, Piaget, and the gestalt psychologists. However, those ideas were too complex to study by observing the behavior of human subjects under controlled conditions. So, because there was no way to confirm or reject such ideas, these potentially more adequate theories found themselves beyond the fringe of what most researchers considered to be the proper domain of science. Today, though, our computers are powerful enough to simulate such artificial minds, provided only that we describe them clearly enough for our programmers to program them.

The first modern computers arrived around 1950, but it wasn't until the 1960s, with the arrival of faster machines and larger memories, that the field of artificial intelligence really began to grow. Many useful systems were invented, and by the late 1970s these were finding applications in many fields. However, even as those applied techniques were spreading, the theoretical progress in that field began to slow, and I found myself wondering what had gone wrong—and what to do about it. The main problem seemed to be that each of our so-called "expert systems" could be used only for some single, specialized application. None of them showed what a person would call general intelligence. None of them showed any signs of having what we call common sense.

For example, some people developed programs that were good at playing games like chess. Other programs were written to prove certain kinds of theorems in mathematics. Others were good at recognizing various kinds of visual patterns—printed characters, for example. But none of the chess programs had any ability to recognize text, nor could the recognition systems prove any theorems, nor could the theorem-proving machine play competently at chess. Some people suggested that we could make our machines more versatile by somehow fusing those programs together into a more integrated whole, but no one had any good ideas about how that could be done. Today, it's still almost impossible to get any two different AI programs to cooperate.

This is precisely the problem I tried to solve in my book *The Society of Mind*. We'd already run into this problem in the 1960s, when Seymour Papert and I started a project to build vision-guided manipulators—that is, robots with eyes and hands. To make a computer that could "see" things, it would need ways to recognize the appearances of objects and the relations between them, but how could you make a computer do such things? Well, at first, each of our researchers tried to develop some particular method. One such project might first try to find the edges of an object and then piece these together into a whole. The trouble was that usually some of the edges were hidden, or were of low optical contrast, or were not really the edges of the object itself but of some decoration on its surface. So edge-finding never worked well enough. Another researcher might try to locate, instead, the surfaces of each object—perhaps by classifying their textures and shadings. This method, too, would sometimes work, but never well enough to be dependable. Eventually, we concluded that no particular method would ever work well enough by itself and we'd have to find ways to combine them.

Why was it so difficult to combine different methods inside the same computer system? Because, I think, our community had never tried to think that way. To do such a thing seemed almost immoral—against the very spirit of "good programming." The accepted paradigm was, and still is, to find a good method for doing the job, and then work on it until you've removed the last bug! Sounds sensible, doesn't it? But eventually we had to conclude that it's a basically wrong idea. After all, even if you did manage to completely debug a program for some particular application, eventually someone would want to use it for some other purpose, in a new environment—and then new bugs would surely appear.

This has been the universal experience with computer programs. In fact, programmers are always joking about this; they often talk about "software rot," which is when a program has been working perfectly for years and then begins to make mistakes, although supposedly nothing has changed. Even today, programmers spend most of their time at trying to make programs work perfectly. The result has been a pervasive trend toward making everything more precise—to make programming into a science instead of an art, by doing everything with perfect logical precision. In my view, this is a misguided idea. What, after all, does it mean for anything to work perfectly? The very idea makes sense only in a rigid, unchanging, completely closed world, like the kinds that theorists make for themselves. Indeed, we can make flawless programs to work on abstract mathematical models based on assumptions that we specify once and then never change. The trouble is that you can't make such assumptions about the real world, because other people are always changing things.

Eventually we made that first robot more robust by installing a variety of methods, but neither it or any of its modern descendants ever were really reliable. We concluded that to make such systems more resourceful and dependable—in a word, more lifelike—we'd better try to understand how human minds manage to rarely get stuck. What are the differences between human thinking and what computers do today? To me, the most striking difference is how almost any error will completely paralyze a typical computer program, whereas a person whose brain has failed at some attempt will find some other way to proceed. We rarely depend upon any one method. We usually know several different ways to do something, so that if one of them fails, there's always another. For example, you can recognize your friends not just by their facial features but by the sound of their voices, by their posture, their gait, their hair color. Given all that variety, we rarely need to perfectly debug any single method. Instead, we learn to recognize the situations in which each of them usually works, and we also learn about conditions in which a method is likely to fail to work. And if all those fail, we can always try to invent a completely new approach.

A century ago, Sigmund Freud was already emphasizing the importance of "negative expertise"—of having knowledge about what *not* to do, a subject that has been entirely ignored by computer scientists and their programmers. Freud talked about censors and other mechanisms that keep us from doing things we've learned to avoid. I suspect that such systems have evolved in our brains, and that a typical brain center is equipped from birth with several different learning mechanisms, some of which accumulate negative knowledge. Thus, one part might accumulate knowledge about when to activate a particular method, another learns to construct "suppressors" or "censors" that can oppose the use of each method, and yet other mechanisms learn what to do when two or more methods come into conflict.

It really is a simple idea—that our minds have collections of different ways to do each of the things they do. Yet this challenges our more common and more ancient ideas about what we are and how we work. In particular, we all share the notion that inside each person there lurks another person, which we call "the self" and which does our thinking and feeling for us: it makes our decisions and plans for us, and later approves, or has regrets. This is much the same idea that Daniel Dennett, who is arguably the best living philosopher of mind, calls the Cartesian Theater—the universal fancy that somewhere deep inside the mind is a certain special central place where all mental events finally come together to be experienced. In that view, all the rest of your brain—all the known mechanisms for perception, memory, language processing, motor control—are mere accessories, which your "self" finds convenient to use for its own inner purposes.

Of course, this is an absurd idea, because it doesn't explain anything. Then why is it so popular? Answer: Precisely because it doesn't explain anything! This is what makes it so useful for everyday life. It helps you stop wondering why you do what you do, and why you feel how you feel. It magically relieves you of both the desire and the responsibility for understanding how you make your decisions. You simply say, "I decided to," and thereby transfer all responsibility to your imaginary inner self. Presumably, each person gets this idea in infancy, from the wonderful insight that you yourself are just another person, very much like the other people you see around you. On the positive side, that insight is profoundly useful in helping you to predict what you, yourself, are likely to do, based on your experience with those others.

The trouble with the single-self concept is that it's an obstacle to developing deeper ideas when we really do need better explanations. Then, when our internal models fail, we're forced to look elsewhere to seek help and advice about what to do in our real lives. Then we find ourselves going to parents, friends, or psychologists, or resorting to those self-help books, or falling into the hands of those folks who claim to have psychic powers. We're forced to look outside ourselves, because that single-self mythology doesn't account for what happens when a person experiences conflicts, confusions, mixed feelings— or for what happens when we enjoy pleasure or suffer pain, or feel confident or insecure, or become depressed or elated, or repelled or infatuated. It provides no clues about why we can sometimes solve problems but other times have trouble understanding things. It doesn't explain the natures of either our intellectual or our emotional reactions—or even why we make that distinction.

What are emotions, anyway? I am developing larger-scale theories of what emotions are, how they work, and how we might learn to control and exploit them. Psychologists have already proposed many smaller theories about different aspects of the mind, but no one since Freud has proposed plausible explanations of how all those systems might interact.

You might ask why this new theory should work, when so many other attempts to explain emotions have failed. My answer is that almost all other such attempts have been looking in the wrong direction. The psychological community suffers from a severe case of physics envy. They've all been searching for some minimal set of basic principles of psychology, some very small collection of amazingly powerful ideas that, all by themselves, can explain how the mind works. They'd like to imitate Isaac Newton, who discovered three simple laws of motion which solved an entire world of problems about mechanics. My method is just the opposite.

The brain's functions simply aren't based on any small set of principles. Instead, they're based on hundreds or perhaps even thousands of them. In other words, I'm saying that each part of the brain is what engineers call a kludge—that is, a jury-rigged solution to a problem, accomplished by adding bits of machinery wherever needed, without any general, overall plan: the result is that the human mind—which is what the brain does—should be regarded as a collection of kludges. The evidence for this is perfectly clear: If you look at the index of any large textbook of neuroscience, you'll see that a human brain has many hundreds of parts—that is, subcomputers—that do different things. Why do our brains need so many parts?

Surely, if our minds were based on only a few basic principles, we wouldn't need so much complexity.

The answer is that our brains didn't evolve in accord with a few well-defined rules and requirements. Instead, we evolved opportunistically, by selecting mutations that favored our survival under the conditions and constraints of many different environments, over the course of at least half a billion years of variation and selection. What, precisely, do all those parts do? We're only beginning to find this out. I suspect that we have much more to learn. When we're all done we'll have found out that many of those mental organs have evolved to correct deficiencies of old ones—that is, deficiencies that did not appear until we got so much smarter. It's characteristic of evolution that after many new structures have developed, it's too late to go back and make much change in those older systems on which we still depend.

In a situation like that, it can be a mistake to focus too much on searching for *basic* principles. More likely, the brain is not based on any such scheme but is, instead, a great jury-rigged combination of many gadgets to do different things, with additional gadgets to correct their deficiencies, and yet more accessories to intercept their various bugs and undesirable interactions—in short, a great mess of assorted mechanisms that barely manage to get the job done.

When I was a kid, I was always compelled to find out how things worked, and I used to dissect all available machinery. I grew up in New York City. My father was an ophthalmologist, an eye surgeon, and our household was always full of interesting friends and visitors—scientists, artists, musicians, and writers. I read all sorts of books, but the ones I loved most were about mathematics, chemistry, physics, and biology. I was never tempted to waste much time at sports, politics, fiction, or gossip, and most of my friends had similar interests. Especially, I was fascinated with the writings of the early masters of science fiction, and I read all the stories of Jules Verne, H.G. Wells, and Hugo Gernsback. Later I discovered the magazines like *Astounding Science Fiction*, and consumed the works of such pioneers as Isaac Asimov, Robert Heinlein, Lester del Rey, Arthur C. Clarke, Harry Harrison, Frederick Pohl, Theodore Sturgeon—as well as the work of their great editor-writer, John Campbell. At first these thinkers were like mythical heroes to me, along with Galileo, Darwin,

Pasteur, and Freud. But there was a difference: all those writers were still alive, and in later years I met them all, and they became good friends of mine, along with their successors—like Gregory Benford, David Brin, and Vernor Vinge, who are fellow scientists as well. What a profound experience it was to be able to collaborate with such marvelous imaginers!

Of course, I also read a great deal of technical literature. But aside from the science fiction, I find it tedious to read any ordinary writing at all. It all seems so conventional and repetitive. To me, the science-fiction writers are our culture's most important original thinkers, while the mainstream writers seem "stuck" to me, rewriting the same plots and subjects, reworking ideas that appeared long ago in Sophocles or Aristophanes, recounting the same observations about human conflicts, attachments, infatuations, and betrayals. Mainstream literature replays again and again all the same old stuff, whereas the science-fiction writers try to imagine what would happen if our technologies and societies—and our minds themselves—were differently composed.

Aside from this, most of my youth was involved in constructing things. Building gadgets. Composing music. Designing new machines. Imagining new processes. When I started at Harvard, in 1946, there was no temptation to play with computers, because there were none. None, that is, except for the Mark I relay computer, which was then being built. I paid almost no attention to it, but one of my Bronx High School of Science classmates, Anthony Oettinger, did. Before long, he became Harvard's first professor of computer science, and around 1952 he wrote one of the first programs to make a computer learn something.

At Harvard, my first concerns were with physics, especially mechanics and optics, and with abstract mathematics, but I soon also got interested in neurophysiology and the psychology of learning. I had the fortune to become attached first to the great B.F. Skinner and then to an extraordinary crew of young professors, including George Miller and Joseph Licklider, who were on the frontier of cybernetics—that great collision between traditional psychology and the new fields of control-engineering that evolved during the Second World War. Perhaps the most important thing that happened, though, was finding a book, *Mathematical Biophysics*, by Nicholas Rashevsky, while browsing through the stacks of science books in Widener Library. Rashevsky showed me how to make abstract models of real things. Then, in Rashevsky's own journal, the *Bulletin of Mathematical Biophysics*, I found the current work of Warren McCulloch and Walter Pitts. First was the original McCulloch and Pitts 1943 paper on threshold neurons and state-machines, which suggested ways to make computerlike machines by interconnecting idealized neurons. Then there was the tremendously imaginative Pitts-McCulloch 1947 paper on vision and group theory, which was the precursor of the group-invariance theorem in *Perceptrons*, the book Seymour Papert and I wrote in 1969. I'm pretty sure that it was works like these, and the flurry of ideas in the early Macy Conference volumes, that kept me thinking about how to make machines that could learn. When Norbert Wiener's revolutionary book, *Cybernetics,* was published in 1949, most of it seemed like old stuff to me, although it taught me a great deal of mathematics.

In the course of thinking about how one might get "neural-network machines" to learn to solve problems, I conceived of what later came to be called Hebb synapses, after the Montreal psychologist Donald Hebb. This inspired me to design a machine in which a randomly connected network of such synapses would compute approximate correlations between stimuli and responses. George Miller got some money from the Air Force Office of Scientific Research, and gave me an account to use to build that machine, which I called the Stochastic Neural Analog Reinforcement Calculator—or SNARC, for short. The machine used about four hundred vacuum tubes and forty little magnetic clutch mechanisms, which would automatically adjust potentiometers, which would in turn control the probabilities that each synapse would transmit a signal from each simulated neuron to another one. The machine worked well enough to simulate a rat learning its way through a maze. I described it in my 1954 Ph.D. thesis, but I don't know how much influence the thesis had on the other researchers. I've never even seen a citation of it, although it has sections proposing other learning mechanisms that so far have never been used.

The SNARC machine was able to do certain kinds of learning, but it also seemed to have various kinds of limitations. It took longer to learn with harder problems, and it sometimes made things worse to use larger networks. For some problems it seemed not to learn at all. This led me to start thinking more about how to solve problems "from the top down," and to start formulating theories about representations and about heuristics for problem solving. In this period, there were only a few people thinking about random neural networks. Nothing very exciting happened in that field—that is, in the field of "general" neural networks, which included looping, time-dependent behavior—until the work of John Hopfield at Caltech, in the early 1980s. There were, however, important advances in the theory of loop-free or "feed forward" networks—notably the discoveries in the late 1950s by Frank Rosenblatt of a foolproof learning algorithm for the machines he called "perceptrons." One novel aspect of Rosenblatt's scheme was to make his machine learn only when correcting mistakes; it received no reward when it did the right thing. This idea has not been adequately appreciated in most of the subsequent work.

The most important other direction in research—of attempting to set down powerful heuristic principles for deliberate, serial problem solving—was already being pursued by Allen Newell, J.C. Shaw, and Herbert Simon. By 1956 they'd developed a system that was able to prove almost all of Russell's and Whitehead's theorems about the field of logic called "proposition calculus." I myself had found a small set of rules that was able to prove many of Euclid's theorems. In the same period, my graduate-school friend John McCarthy was making progress in finding logical formulations for a variety of commonsense reasoning concepts.

Soon the field of artificial intelligence began to make rapid progress, with the spectacular work of Larry Roberts on computer vision, and the work of Jim Slagle on symbolic calculus, and around 1963, ARPA—the Defense Department's Advanced Research Projects Agency—began to support several such laboratories on a reasonably generous scale. Rosenblatt's neural-network followers were also making ambitious proposals, and this led to a certain amount of polarization. This was partly because some of the neural-network enthusiasts were actually pleased with the idea that they didn't understand how their machines accomplished what they did. When Seymour and I managed to discover some of the reasons why those machines could solve certain problems but not others, many of those neovitalists interpreted this not as a mathematical contribution but as a political attack on their work. This evolved into a strange mythology about the nature of our research—but that's another story.

What can you do when you have a problem that you can't seem to solve in a single step? Then you have to find ways to break it up into subproblems and try to find ways to solve each of those. By the end of the 1960s, quite a few people were thinking about this, and I tried to pull the field together and write a book about it. The trouble was that we were discovering new methods faster than we could write them down, so in 1961 I pulled together as much as I could and published one very large paper titled "Steps Toward Artificial Intelligence." Although this was not strictly a synthesis, it did establish a fairly uniform terminology for the field, and establish the subject as a well-defined scientific enterprise. Some techniques proposed in that paper have still not yet been adequately explored.

After pursuing several different approaches to the problem of making artificial intelligence—and trying to decide which method might be best—I finally realized that there's no best way. Each particular method has advantages for particular kinds of situations. This means that they key to making a smart machine is inventing ways to *manage* a variety of resources—and this led to what Seymour and I called the society-of-mind theory. If you look at the brain, you see that there are hundreds of different kinds of neural nets there—hundreds of different kinds of structures. When you injure different pieces of the brain, you see different symptoms. That led to the idea that maybe you can't understand anything unless you understand it in several different ways, and that the search for the single truth—the pure, best way to represent knowledge—is wrongheaded.

The reason it's wrongheaded is that if you understand something in just one way, and the world changes a little bit and that way no longer works, you're stuck, you have nowhere to go. But if you have three or four different ways of representing the thing, then it would be very hard to find an environmental change that would knock them all up. People are always getting into situations that are a little bit different from old ones. You have to accumulate different viewpoints and different ways of doing things and different mechanisms. If you want to do learning with neural nets, you can't just use one kind of neural net; you'll probably have to design different types suited for remembering stories, for representing geometrical structures, for causal interactions, for making chains-of-reasoning steps, for the semantic relations of language expressions, for the sorts of two-dimensional representations needed for vision, and so on. The secret of intelligence is that there is no secret—no special, magical trick.

ROGER SCHANK: Marvin Minsky is the smartest person I've ever known. He's absolutely full of ideas, and he hasn't gotten one step slower or one step dumber. One of the things about Marvin that's really fantastic is that he never got too old. He's wonderfully childlike. I think that's a major factor explaining why he's such a good thinker. There are aspects of him I'd like to pattern myself after. Because what happens to some scientists is that they get full of their power and importance, and they lose track of how to think brilliant thoughts. That's never happened to Marvin.

Marvin should have been my thesis advisor. I wouldn't say that I'm his student, but I appreciate everything he does. His point of view is my point of view. I like his ideas, and he likes mine. They're similar. I especially resonate with his idea that the mind is a collection of kludges. We see the world in the same way.

STEVEN PINKER: Marvin is a brilliant man and deserves a lot of credit for getting cognitive science started, and also for posing the question of psychology as reverse software engineering. I admire him for having put the question that way. But Marvin's ideas are more hit-and-miss when he broadcasts opinions on the structure of the mind based on his own intuitions. He's become a guru, as opposed to following in depth the actual empirical work done in the laboratory on specific aspects of mind: the work on how people see, how people talk, how languages differ, what the logic underlying language is. Too much of his work comes out of armchair pronouncements, as opposed to fieldwork and laboratory work.

FRANCISCO VARELA: Minsky has some very interesting insights, such as his notion of the society of mind. As a scientist, he's one of the many voices of people coming up with a fundamental insight about multiple levels of agents in the cognitive system.

As a person, I find him a pain in the ass, but that's on a different register. What he does is interesting. My run-ins with him are based on two things. One is, as I said, that he's a pain in the ass, an arrogant son of a bitch. Minsky is one of those people who's very quick on the draw without knowing the other person's work. I know this from personal experience; it happened at a conference we both attended. Before I could open my mouth, he already had an idea about me. He actually, at one point, said, "The reason I'm angry with you is because I'm angry with Fernando Flores and Terry Winograd." He obviously loves Winograd, so he was angry with me, as a Chilean and a friend of Flores', because Flores, according to Minsky, had led Winograd away from Minsky's clear path of AI into what Minsky considered some kind of bullshit. That might or might not be true, but in any case, in his confusion, he put me in the same boat. It's not very interesting, quite frankly.

RICHARD DAWKINS: I respect Marvin Minsky as the founding father of artificial intelligence, which is a subject that has fascinated me. I've been on the periphery of it, but I've never actually done any work on it. I've attended the occasional conference on it. I know him only as a father figure in that field.

DANIEL C. DENNETT: Minsky and I have very similar views on a lot of things; it's just that I think the problems are harder than Marvin does. Marvin thinks he's solved them all, and he expects everyone to understand that. It's not as hard as they think, he thinks, if they'd all just pay attention to him! He has the basic solutions; he just has to work out some of the details. Curiously, even if that's correct—and by some stretch of the imagination it might be correct—Minsky's presumption is wrong. Even if Minsky is right, you can't get everything you need just by reading Minsky. His writing is too compressed. It makes Minsky into a philosopher; he'll probably hate me for saying this! There's nothing more common to scholars of philosophy than to discover that they can read later developments into the chance remarks, or the few well-chosen remarks, of some earlier philosopher. Everything that's ever been done in philosophy is just footnotes to Plato, that sort of thing. If you operate at a certain level of abstraction, that can be true without being very exciting. It can be true because if you ignore enough of the details and have basically good instincts about where the truth lies, you can probably manage to say nothing but the truth. In one sense, you get it right, but you don't get enough of it right to help.

Marvin has done wonderful things. One of his main contributions has been showing again and again just how much you can build with the simple building blocks of artificial intelligence. If you're imaginative, and if you have a certain weird discipline whereby you don't hide the hard problems but try to address them, you'll find him inspiring, because he shows how simple ideas can have big families of interesting consequences. That's what artificial intelligence is all about. The cost is that you get a few dead ends, too. Not all of Minsky's ideas are good, but many of them are.

Chapter 9

ROGER SCHANK

"Information Is Surprises"

MARVIN MINSKY: Roger Schank has pioneered many important ideas about how knowledge might be represented in the human mind. In the early 1970s, he developed a concept of semantics that he called "conceptual dependency," which plays an important role in my book The Society of Mind. He's also developed other paradigms, involving representing knowledge in various types of networks, scripts, and storylike forms.

• • •

ROGER SCHANK is a computer scientist and cognitive psychologist; director of the Institute for the Learning Sciences, at Northwestern University; John Evans Professor of Electrical Engineering and Computer Science, and professor of psychology and of education and social policy; author of fourteen books on creativity, learning, and artificial intelligence, including The Creative Attitude: Learning to Ask and Answer the Right Questions, *with Peter Childers (1988),* Dynamic Memory *(1982),* Tell Me A Story *(1990), and* The Connoisseur's Guide to the Mind *(1991).*

ROGER SCHANK: My work is about trying to understand the nature of the human mind. In particular, I'm interested in building models of the human mind on the computer, and especially working on learning, memory, and natural-language processing. I'm interested in how people understand sentences, how they remember things, how they get reminded of one event by another, and how they learn from one experience and use it to help them in other events. Most people in the field associate me with the idea that there are mental structures called "scripts," which help you understand a sequence of events and allow you to make inferences from those events—inferences that essentially guide your plans or behavior through those events.

Information is surprises. We all expect the world to work out in certain ways, but when it does, we're bored. What makes something worth knowing is organized around the concept of expectation failure. Scripts are interesting not when they work but when they fail. When the waiter doesn't come over with the food, you have to figure out why; when the food is bad or the food is extraordinarily good, you want to figure out why. You learn something when things don't turn out the way you expected.

The most important thing to understand about the mind is that it's a learning device. We're constantly trying to learn things. When people say they're bored, what they mean is that there's nothing to learn. They get unbored fast when there's something to learn. The important thing about learning is that you can learn only at a level slightly above where you are. You have to be prepared.

My most interesting invention is probably my theory of MOPs and

TOPs—memory-organization packets and theme-organization packets—which is basically about how human memory is organized: any experience you have in life is organized by some kind of conceptual index that's a characterization of the important points of the experience. What I've been trying to do is understand how memory constantly reorganizes, and I've been building things called dynamic memories. My most important work is the attempt to get computers to be reminded the way people are reminded.

I also made early contributions to the field of natural-language processing, where I went head to head with the linguists, who were working on essentially syntactical models of natural language. I was interested in conceptual models of natural language. I was interested in the question of how, when you understand a sentence, you extract meaning from that sentence independent of language.

I've gotten into lots of arguments with linguists who thought that the important question about language was its syntactic structure, its formal properties. I'm what is often referred to in the literature as "the scruffy"; I'm interested in all the complicated, un-neat phenomena there are in the human mind. I believe that the mind is a hodgepodge of a range of oddball things that cause us to be intelligent, rather than the opposing view, which is that somehow there are neat, formal principles of intelligence.

An example I used in my book *Dynamic Memory* is the case of the steak and the haircut. The story is that I was complaining to a friend that my wife didn't cook steak the way I liked it—she always overcooked it. My friend said, "Well, that reminds me of the time I couldn't get my hair cut as short as I wanted it, thirty years ago in England." The question I ask is, How does such reminding happen and why does it happen? The "how" is obvious. What are the connections between the steak and the haircut? If you look at it on a conceptual level, there's an identical index match: we each asked somebody who had agreed to be in a service position to perform that service, and they didn't do it the way we wanted it. There are a number of questions you can ask. First, how do we construct such indices? Obviously, my friend constructed such an index in order to find, in his own mind, the story that had the same label. Second, why do you construct them? And the answer is that you're trying to understand the universe and you need to match incoming events to past experiences. This is something I call "case-based reasoning." The idea that you would then make that match obviously has a purpose. It's not hard to understand what the purpose would be; the purpose is learning. Because how would you learn from new experiences otherwise?

The case-based-reasoning model says you process a new experience by constructing some very abstract label for it, and that label is the index into memory. Many things in memory have been labeled that way; you find them and you make comparisons, almost like a scientist, between the old experience and the new experience, to see what you can learn from the old experience to help you understand the new experience. When you finish that process, you can go back into your mind and add something that will help fix things. For example, I can imagine my friend saying, "Well, I guess that experience I had in England wasn't so unusual; there really are a lot of times when people don't do things because they think it would be too extreme." Sure enough, I go back and check with my wife, and the reason she overcooks the steak is that she thinks I want it too rare.

One of the problems we've had in AI is that in the early years—in the sixties and seventies—you could build programs that seemed pretty exciting. You could get a program to understand a sentence, or translate a sentence. Twenty years later, it's not exciting any more. You've got to build something real, and in order to build real things you have to work with real problems. Understanding how learning might take place when people are telling stories to each other; understanding how somebody might produce a sentence, or how somebody might make an inference, or how somebody might make an explanation: those kinds of things have interested me, whereas in AI your average person was much more interested in the formal properties of vision, say, or building robotic systems, or proving theorems, or things that are more logically based.

What I've learned in twenty years of work on artificial intelligence is that artificial intelligence is very hard. This may sound like a strange thing to say, but there's a sense in which you have only so many years to live, and if we're going to build very intelligent machines it may take a lot longer to do than I personally have left in life. The issue is that machines have to have a tremendous amount of knowledge, a tremendous amount of memory; the software-engineering problems are phenomenal. I'm still interested in AI as a theoretic enterprise—I'm as interested in cognitive science and the mind as I've ever been—but since I'm a computer scientist, I like to build things that work.

One thing that's clear to me about artificial intelligence and that, curiously, a lot of the world doesn't understand is that if you're interested in intelligent entities there are no shortcuts. Everyone in the AI business, and everyone who is a viewer of AI, thinks there are going to be shortcuts. I call it the magic-bullet theory: somebody will invent a magic bullet in the garage and put it into a computer, and Presto! the computer's going to be intelligent. Journalists believe this. There are workers in AI who believe it, too; they're constantly looking for the magic bullet. But we became intelligent entities by painstakingly learning what we know, grinding it out over time. Learning about the world for a ten-year-old child is an arduous process. When you talk about how to get a machine to be intelligent, what it has to do is slowly accumulate information, and each new piece of information has to be lovingly handled in relation to the pieces already in there. Every step has to follow from every other step; everything has to be put in the right place, after the previous piece of information. If you want to get a machine to be smart, you're going to have to put into it all the facts it may need; this is the only way to give it the necessary information. It's not going to mysteriously acquire such information on its own.

You can build learning machines, and the learning machine could painstakingly try to learn, but how would it learn? It would have to read the *New York Times* every day. It would have to ask questions. Have conversations. The concept that machines will be intelligent without that is dead wrong. People are set up to be capable of endless information accumulation and indexing; finding information and connecting it to the next piece of information—that's all anyone is doing.

One of the most interesting issues to me today is education. I want to know how to rebuild the school system. One thing is to look at how people learn, right now, and how the schools work, right now, and see if there's any confluence. In schools today, students are made to read a lot of stuff, and they're lectured on it. Or maybe they see a movie. Then they do endless problems, then they get a multiple-choice test of a hundred questions. The schools are saying, "Memorize all this. We're going to teach you how to memorize. Practice it, we'll drill you on it, and then we're going to test you."

Imagine that this is how I'm going to teach you about food and wine. We're going to read about food and wine, and then I'll show you films about food and wine, and then I'll let you solve problems about the nature of food and wine, like how to decant a bottle of wine, what the optimal color is for a Bordeaux, and so forth. And then I'll give you a test.

Would you learn to appreciate food and wine this way? Would you learn *anything* about food and wine? The answer is no. Because what you have to do to learn about food and wine is eat and drink. Memorizing all the rules, or discussing the principles of cooking, isn't going to do any good if you don't eat and drink. In fact, it works the other way around. If you eat and drink a lot, I can get you interested in those subjects. Otherwise I can't.

Everything they teach in school is oriented so that they can test it to show that you know it, instead of taking note of the obvious, which is that people learn by doing what people want to do. The more they do, the more curious they get about how to do it better—*if* they're interested in doing it in the first place. You wouldn't teach a kid to drive by giving him the New York State test manual. If you want to learn how to drive, you have to drive a lot. Most schools do everything but allow kids to experience life. If kids want to learn about what goes on in the real world, they have to go out into the real world, play some role in it, and have that motivate learning. Errors in learning by doing bring out questions, and questions bring out answers.

What kids learn in high school or college is antilearning. By reading Dickens in ninth grade, I learned to hate Dickens. Ten years later, I picked up Dickens and it was interesting, because I was ready to read it. What I learned in high school was something useless—that Dickens is awful. A ninth-grade kid isn't ready for this. Why do they teach it? Because in the nineteenth century that was the literature of the time, and that's when they designed the curriculum still used in practically all schools today.

I don't think there should be a curriculum. What kids should do is follow the interests they have, with an educated advisor available to answer their questions and guide them to topics that follow from the original interest. Wherever you start, you can go somewhere else naturally. The problem is that schools want everyone to be in lockstep: everyone has to learn this on this day and that on that day. School is a wonderful baby-sitter. It lets the parents go to work and keeps the kids from killing each other.

Learning takes place outside of school, not in school, and kids who want to know something have to find out for themselves by asking questions, by finding sources of material, and by discounting anything they learned in school as being irrelevant.

Most teachers feel threatened by questions. Obviously, good teachers love to hear good questions, but the demographics don't allow them to answer all the questions anyway. This is where computers can come in. One-on-one teaching is what matters. In the old days, rich people hired tutors for their kids. The kids had one-on-one teaching, and it worked. Computers are the potential savior of the school system, because they allow one-on-one teaching. Unfortunately, every piece of educational software you see on the market today is stupid, because it was designed to follow the same old curriculum.

At the Institute for the Learning Sciences, at Northwestern, we designed a new computer program to teach biology, in which you get to design your own animal. The National Science Foundation said that this program wouldn't fit into the curriculum, because biology isn't taught in the sixth grade, which is the level at which the program works. Furthermore, since each kid would have a different conversation with the computer, how could tests be given on what was learned?

The real problem is the idea that knowledge is represented as a set of facts. It's not. You might want to know those facts, but it's not the knowing of the facts that's important. It's how you got that knowledge, the things you picked up on the way to getting that knowledge, what motivated the learning of that knowledge. Otherwise what you're learning is just an unrelated set of facts. Knowledge is an integrated phenomenon; every piece of knowledge depends on every other one. School has to be completely redesigned in order to be able to make this happen.

This is where the computer comes in, through computer programs that are knowledgeable and can have conversations with kids about whatever subject the kids want to talk about. Kids can begin to have conversations about biology or history or whatever, and have their interest sustained. What you need are computer programs that can do the kind of one-on-one teaching that a good teacher could do if he or she had the time to do it.

Not long ago, to prepare for a conference, I read Darwin. Doing this reaffirmed my belief in not reading, because if I had read Darwin at any other time in my life I wouldn't have understood him. I was only capable of understanding Darwin in a meaningful way by reading him this time, because I understood something about what his argument was with respect to arguments I was trying to make. I could internalize it. Darwin's very clever. He said all kinds of interesting things that I wouldn't have regarded as relevant twenty years ago.

The issue is reading when you're prepared to read something. For instance, at this moment I'm not thinking about consciousness, so if I read Dan Dennett, he would do one of two things to me. He would cause me to react to his thinking about consciousness, which means that I would forever think about consciousness in his metaphor. This is useless to me, if I want to be creative. Secondly, I would reject his theories out of hand and find the book and the subject not worth thinking about. This also is bad. I don't see the point of reading his book unless at this moment I've thought about consciousness and am prepared to see what he thinks. That's my view of reading. The problem is that intellectuals say to each other, "Oh my God, haven't you read X?" It's academic one-upmanship.

The MIT linguist Noam Chomsky represents everything that's bad about academics. He was my serious enemy. It was such an emotional topic for me twenty years ago that at one point I couldn't even talk about it without getting angry. I'm not sure I'm over that. I don't like his intolerant attitude or what I consider tactics that are nothing less than intellectual dirty tricks. Chomsky was the great hero of linguistics. In his view of language, the core of the issue is syntax. Linguistics is all about the study of syntax. Language should be looked at in terms of Chomsky's notion of its "deep structure." Part of Chomsky's cleverness in referring to deep structure was to use these wonderful words in a way that everyone assumed to be something other than what he meant.

What Chomsky meant by "deep structure" was that you didn't have to look at the surface structure of a sentence—the nouns and the verbs, and so forth. But what any rational human being would have thought he meant by "deep structure," he emphatically did not mean. You would imagine that a deep structure would refer to the ideas behind the sentence, the meaning behind the sentence. But Chomsky stopped people from working on meaning.

I was sufficiently out of that world so that I could yell and scream and say that meaning is the core of language. I went through every point he ever made, and made fun of each one. He was always an easy target, but he had a cadre of religious academic zealots behind him who essentially would listen to no one else.

Here's an example of an argument I might have had with him in the late sixties. The sentence "John likes books" means that John likes to read. "Oh no," Chomsky might say, "John has a relationship of liking with respect to books, but he might not like to read."

Part of what linguistic understanding is about is understanding meaning: what you can assume to be absolutely true, and what you can assume to be true some of the time, or likely to be true. I call this inference. But Chomsky would say, "No, inference has nothing to do with language, it has to do with memory, and memory has nothing to do with language."

That comment is totally absurd. The psychology of language is what's at issue here. Meaning, inferences, and memory are a very deep part of language. Chomsky explicitly states in his most important book, *Aspects of the Theory of Syntax*, that memory is not a part of language and that language should be studied in the abstract. Language, for Chomsky, is a formal study, the study of the mathematics of language. I can see someone making arguments about language from a perspective of mathematical theory, but not if you are a founding member of the editorial board of *Cognitive Psychology,* and not if legions of psychologists are writing articles and conducting experiments based upon your work. Chomsky tried to have it both ways.

In Chomsky's view, the mind should behave according to certain organized principles, otherwise he wouldn't want to study it. I don't share that view. I'll study the mind, and whatever I get is O.K. Let it all be mud. Fine, if that's what it is. There are many scientists who'd like the mind to be scientific. If it isn't scientific—neat and mathematical—they don't want to have to deal with it. Chomsky has always adopted the physicist's philosophy of science, which is that you have hypotheses you check out, and that you could be wrong. This is absolutely antithetical to the AI philosophy of science, which is much more like the way a biologist looks at the world. The biologist's philosophy of science says that human beings are what they are, you find what you find, you try to understand it, categorize it, name it, and organize it. If you build a model and it doesn't work quite right, you have to fix it. It's much more of a "discovery" view of the world, and that's why the AI people and the linguistics people haven't gotten along. AI isn't physics.

MURRAY GELL-MANN: I know Roger Schank slightly, and I find that his work has many appealing characteristics. Working with the concept of scripts, he was led into a huge project in education, using computers. As I listened to his description of some of the ideas behind the project, I found myself in sympathy with many of them.

Ever since teaching machines of the most primitive kind were first invented, I have thought that computers, programmed intelligently to function as teaching machines, could be used most effectively for education, because they would allow students to go through the routine parts of learning without using up teachers' time and without subjecting the student to the embarrassment of public viewing of his or her preliminary answers to questions. As we know, there is not really such a thing as education. There is only helping somebody to learn, and the learning process is a complex adaptive system: fooling around, making mistakes, somehow having contact with reality or truth, correcting the mistakes, assuring self-consistency, and so on. You can go through that with a machine without being subjected to ridicule. At the same time, the machine can keep track of your thought processes if necessary. When certain thought processes are in error, the machine can tell you that, so that you can change them. Furthermore, the people relieved of the necessity of doing the routine jobs carried out by the teaching machine can be saved for other duties—ones that really require a human being.

I've always thought that university education, including full-scale lecture courses covering the ground of well-known subjects on which excellent books have been published, are simply an illustration of how the universities have failed to adapt, after five hundred years, to the invention of printing. For those who prefer to learn by listening and watching, videotaped courses by some of the best lecturers in the world are now—or may soon be—available. Presumably universities will adapt slowly to such modern inventions as well. In medieval times, books were published by having a *lector* read his manuscript to a roomful of *scriptores*, who wrote it down. Many of the students at the university—say, in theology—were too poor to buy books produced by this expensive method, and so at the university a theology professor would read his book to the students, who would act as their own *scriptores* and write down what the teacher said.

With the invention of printing, this system became obsolete, but the universities have still not noticed that, after more than five hundred years. Of course, a lecture can serve very important purposes: It can convey brand-new information, along with the exciting character of that information. A dramatic lecture can serve to present the speaker as a role model to the people in the audience. I have nothing against the occasional lecture. But the idea that at each college and university some professor has to give a series of lectures covering the ground of a subject such as electromagnetic theory seems totally insane to me. If professors really want to assist learning, they can answer questions when students are stuck, assign challenging problems and fascinating reading, and give occasional exciting talks. And of course they can choose textbooks, and if necessary, series of videotaped lectures. In brief, they can serve as resources for students engaged in the complex adaptive learning process.

MARVIN MINSKY: Roger Schank has pioneered many important ideas about how knowledge might be represented in the human mind. In the early 1970s, he developed a concept of semantics that he called "conceptual dependency," which plays an important role in my book *The Society of Mind*. He's also developed other paradigms, involving representing knowledge in various types of networks, scripts, and storylike forms. Each of these ideas suggests, in turn, another new theory of memory. In this way, Schank has been enormously productive in the artificial-intelligence field. He's changed his focus from year to year, so that in each of several different periods he would train a new generation of students in different theories. Then he would force them to build computer models of those theories, so that the rest of us could see for ourselves what these models could and could not do. Most of the models were based on novel ways to represent the meanings of verbal expressions.

Ironically, Schank has been opposed and almost persecuted by the language theorist Noam Chomsky, who himself generated several families of new ideas. Generally, Chomsky both ridiculed Schank's approach—sometimes by saying curtly that it just wasn't interesting—and completely ignored the significance of Schank's results. I used the word "ironically" because the work of Schank and Chomsky is so strikingly complementary. Chomsky seems almost entirely concerned with the formal syntax of sentences, to the nearly total exclusion of how words are actually used to represent and communicate ideas from one person to another. He thus ignores any models indicating that syntax is only an accessory to language. For example, no one has any trouble in understanding the story implied by the three-word utterance "thief, careless, prison," although it uses no syntax at all. Schank and his students, however, have demonstrated several ways to deal with such intricate meanings. It was quite hard to persuade our colleagues to consider these kinds of theories. Sometimes, it seems, the only way to get their attention is by shocking them. Roger Schank is good at this. His original discussion of conceptual dependency used such examples as "Jack threatened to choke Mary unless she would give him her book." His technical representation of this idea is that Jack transfers into Mary's mind the conceptualization that if she doesn't transfer the possession of the book to him, he'll cut off her windpipe, so that she won't get enough air to live. I once asked Roger why so many of his examples were so bloodthirsty. He replied, "Ah, but notice how clearly you remember them!"

FRANCISCO VARELA: Roger Schank is somebody I don't know personally, but I know what he has to say. In some sense, Schank is another good example of somebody who stands on the opposite side of the fence from me, regarding the understanding of mind. For Schank, there's a fundamental assumption of mind as some kind of logical machine—rationalist mind. The basic approach I take is that this is just simply a hangover from Western tradition, and that "mind" is fundamentally not rational. It's not a decision-making, software-type process. In that sense, Schank serves as a sparring partner.

STEVEN PINKER: Roger Schank, I think, is another example of how a scientist's reaction to a theory is "What have you done for me lately?" Roger is more of an engineer than a scientist; he doesn't have a theory of how the human brain learns and uses language based on detailed study of children and adults talking. His goal was to build computer programs that understand, and that's a very different enterprise. There was a rather acrimonious debate between him and Chomsky's followers in the 1970s. But a lot of that energy may have been wasted, because they were talking past each other.

Chomsky was looking at one small part of the problem of understanding human language—namely, how children acquire the grammar of their mother tongue. His answer was one that I agree with: that the brain has, among other things, some circuitry dedicated to learning grammar, and some aspects of the design of grammar are built in. Chomsky argued that this is one of the most interesting questions about language, but he'd be the first to admit that it's only a small part of the scientific problem of how people use language in understanding stories and in conversation—to say nothing of the problem of what the best ways are of building a computer to do these things. Roger had a much more ambitious goal, in terms of engineering—that is, to write programs that could understand stories. He said, "A theory like Chomsky's doesn't help me solve my problem; knowing the universal constraints on grammars of all languages isn't going to help me devise a program that can understand stories in English. Therefore Chomsky was wrong about language."

This was unfortunate. Much of the debate between Chomsky and Schank is another case of the blind men and the elephant. They're asking different questions, so the answers they come up with aren't really contradictory. Chomsky, in my opinion, is right in saying that there's an autonomous mental organ for grammar and that a child can acquire grammar only if the basic design of the grammar of the world's languages is in some sense built in. Roger is right in that actual use of language, in conversation or understanding, involves a lot more than grammar—such as knowledge of how people interact with one another in typical situations—and that therefore to tell the whole story about how conversation works, you can't simply have a theory of grammar but you must embed it in a theory of knowledge about the world and social interactions.

W. DANIEL HILLIS: The Roger Schank I knew was a thorn in everybody's side—constructively so. The interesting thing about Roger Schank, something he shares with Minsky, is the fact that he's produced an incredible string of students. Anybody who's produced such a great string of students has to be a constructive pain in the ass. He's always taken an adversarial stance in his theories. He doesn't just say, "Here's my theory." He says, "Here's why I'm right and everybody else is an idiot." He's often right.

DANIEL C. DENNETT: I've always relished Schank's role as a gadfly and as a naysayer, a guerrilla in the realm of cognitive science, always asking big questions, always willing to discard his own earlier efforts and say they were radically incomplete for interesting reasons. Part of Roger's view is that the mind is an amazing collection of gadgets, held together with some very interesting sorts of baling wire. With that sort of view, of course, you can't have a systematic scientific research program, so he doesn't try to. He's an opportunistic explorer of his own ideas. He still gets interesting results. A lot of his effort is spent trying to lead people in what he thinks are the right directions and fomenting whatever revolution he's currently fomenting, rather than trying to work out in a solitary way the final truth about anything. He's a gadfly and a good one.

One of the ideas he's best known for is "scripts"—stereotypic situation-types or narrative fragments out of which, he claimed, we construct most if not all of our cognitive prowess. He probably would agree that now his own efforts on behalf of scripts can be seen as the unintended refutation of a superficially promising idea. There was something right about it, but everybody started beating up on the idea, and the more we looked, and the more we saw what was involved if you tried to make it work, the more we could see that scripts by themselves couldn't do the job he first thought they could. Roger himself was probably the most insightful critic of scripts. Good! We learned something that wasn't obvious. People who say that it was obvious from the outset reveal that they haven't thought very seriously about the problem. It wasn't clear what scripts could and couldn't do until Roger forced us to look hard at the idea.

Chapter 10

DANIEL C. DENNETT

"Intuition Pumps"

MARVIN MINSKY: *Dan Dennett is our best current philosopher. He is the next Bertrand Russell. Unlike traditional philosophers, Dan is a student of neuroscience, linguistics, artificial intelligence, computer science, and psychology. He's redefining and reforming the role of the philosopher. Of course, Dan doesn't understand my society-of-mind theory, but nobody's perfect.*

• • •

DANIEL C. DENNETT is a philosopher; director of the Center for Cognitive Studies and Distinguished Arts and Sciences Professor at Tufts University; author of Content and Consciousness *(1969)*, Brainstorms *(1978)*, Elbow Room: The Varieties of Free Will Worth Wanting *(1984)*, The Intentional Stance *(1987)*, Consciousness Explained *(1991)*, and coauthor *(with Douglas R. Hofstadter) of* The Mind's I *(1981)*.

DANIEL C. DENNETT: If you look at the history of philosophy, you see that all the great and influential stuff has been technically full of holes but utterly memorable and vivid. They are what I call "intuition pumps"—lovely thought experiments. Like Plato's cave, and Descartes's evil demon, and Hobbes' vision of the state of nature and the social contract, and even Kant's idea of the categorical imperative. I don't know of any philosopher who thinks any one of those is a logically sound argument for anything. But they're wonderful imagination grabbers, jungle gyms for the imagination. They structure the way you think about a problem. These are the real legacy of the history of philosophy. A lot of philosophers have forgotten that, but I like to make intuition pumps.

I like to think I'm drifting back to what philosophy used to be, which has been forgotten in many quarters in philosophy during the last thirty or forty years, when philosophy has become a sometimes ridiculously technical and dry, logic-chopping subject for a lot of people—applied logic, applied mathematics. There's always a place for that, but it's nowhere near as big a place as a lot of people think.

I coined the term "intuition pump," and its first use was derogatory. I applied it to John Searle's "Chinese room," which I said was not a proper argument but just an intuition pump. I went on to say that intuition pumps are fine if they're used correctly, but they can also be misused. They're not arguments, they're stories. Instead of having a conclusion, they pump an intuition. They get you to say "Aha! Oh, I get it!"

The idea of consciousness as a virtual machine is a nice intuition pump. It takes a while to set up, because a lot of the jargon of artificial intelligence and computer science is unfamiliar to philosophers or other people. But if you have the patience to set some of these ideas up, then you can say, "Hey! Try thinking about the idea that what we have in our heads is software. It's a virtual machine, in the same way that a word processor is a virtual machine." Suddenly, bells start ringing, and people start seeing things from a slightly different perspective.

Among the most appealing ideas in artificial intelligence are the variations on Oliver Selfridge's original Pandemonium idea. Way back in the earliest days of AI, he did a lovely program called Pandemonium, which was very well named, because it was a bunch of demons. Pan-demonium. In his system, there were a lot of semi-independent demons, and when a problem arose, they would all jump up and down and say, in effect: "Me! me! me! Let me do it! I can do it!" There would be a brief struggle, and one of them would win and would get to tackle the problem. If it didn't work, then other demons could take over.

In a way, that was the first connectionist program. Ever since then, there have been waves of enthusiasm in AI for what are, ultimately, evolutionary models. Connectionist models are ultimately evolutionary. They involve the evolution of connection strengths over time. You get lots of things happening in parallel, and what's important about them is that, from a Calvinist perspective, they look wasteful. They look like a crazy way to build anything, because there are all these different demons working on their own little projects; they start building things and then they tear them apart. It seems to be very wasteful. It's also a great way of getting something really good built—to have lots of building going on in a semicontrolled way, and then have a competition to see which one makes it through to the finals.

The AI researcher Douglas Hofstadter's jumbo architecture is a very nice model that exhibits those features. The physicist Stephen Wolfram has some nice models, although they're not considered AI. These architectures are very different from good old-fashioned AI models, which, you might say, were bureaucratic, with a chain of command and a boss and a sub-boss and a bunch of sub-sub-bosses, and delegation of responsibility, and no waste. Hofstadter once commented that the trouble with those models is that the job descriptions don't leave room for fooling around. There aren't any featherbedders. There aren't any people just sitting around, or making trouble. Mother Nature doesn't design things that way. When Mother Nature designs a system, it's "the more the merrier, let's all have a big party, and somehow, we'll build this thing." That's a very different organizational structure. My task, in a way, is to show how, if you impose those ideas—of a plethora of semi-independent agents acting in an only partly organized way with lots of "waste motion"— on the brain, all sorts of things begin to fall into place, and you get a different view of consciousness.

As technology changes, we change. As computers evolve, our philosophical approach to thinking about the brain will evolve. In the history of thinking about the brain, as each new technology has come along it's been enthusiastically exploited: clockwork and wires and pulleys back in Descartes's day, then steam engines and dynamos and electricity came in, and then the telephone switchboard. We should go back earlier. The most pervasive of all of the technological metaphors people have used to explain what goes on in the brain is writing—the idea that we think about the things happening in the brain as signals, as messages being passed. You don't have to think about telegraphy or telephones, you just have to think about writing messages.

The idea that memory is a storehouse of things written is already a metaphor, and even a bad metaphor. The very idea that there has to be a language of thought doesn't make sense unless you think of it as a written language of thought. A spoken language of thought won't get you much of anything. One of the themes that interests me is the idea of talking before you know you're talking, before you know what talking is, which we all do. Children do it. There's a big difference between talking and self-conscious talking, which, if you get clear about it, helps with the theory of language.

People couldn't think of the brain as a storehouse at all before there was a written language. There wasn't a mind/body problem, and there weren't any theories of mind, even if you go back to the ancient Greeks, even Plato and Aristotle. You find nothing much in the way of what looks like theorizing about this. What they did say was rather bad.

The basic idea of computation, as formulated by the mathematicians John von Neumann and Alan Turing, is in a class by itself as a breakthrough idea. It's the only idea that begins to eliminate the middleman. What was wrong with the telephone-switchboard idea of consciousness was that you have these wires that connect what's going on out at the eyeballs into some sort of control panel. But then you still have that clever homunculus sitting at the control panel doing all the work.

If you go back further, David Hume theorized about impressions and ideas. Impressions were like slides in a slide show, and ideas were faint copies—poor-quality Xerox copies—of the original pictures. He tried to dream up a chemistry theory, a phony theory of valences which would suggest how one idea could bring the next one along. I explained this idea to a student one day who said that Hume was trying to get the ideas to think for themselves. That's exactly what Hume was trying to do. He was trying to get rid of the thinker, because he realized that that was a dead end. If you still have that middleman in there doing all the work, you haven't made any progress. Hume's idea was to put little valence bonds between the ideas, so that each one could think itself and then get the next one to think itself, and so forth—getting rid of the middleman. But it didn't work.

The only idea anyone has ever had which demonstrably does get rid of the middleman is the idea of computers. Homunculi are now O.K., because we know how to discharge them. We know how to take a homunculus and break it down into smaller and smaller homunculi, eventually getting down to a homunculus that you can easily replace with a machine. We've opened up a huge space of designs—not just von Neumannesque, old-fashioned computer designs but the designs of artificial life, the massively parallel designs.

Right now I'm working on how you get rid of the Central Meaner, which is one of the worst homunculi. The Central Meaner is the one who does the meaning. Suppose I say, "Please repeat the following sentence in a loud clear voice: 'Life has no meaning, and I'm thinking of killing myself.'" You might say it, but I don't think you'd mean it, because—some people would be tempted to say—even though your body uttered the words, your Central Meaner wasn't endorsing it, wasn't in there saying, in effect, "This is a real speech act. I mean it!"

I've recently been looking at the literature on psycholinguistics, and sure enough, they have a terrible time dealing with production of speech. All their theories are about how people understand speech, how they comprehend it, how they take it in. But there isn't much at all about how people generate speech. If you look at the best model that anyone's come up with to date, the Dutch psycholinguist Willem Levelt's model, he's got a "blueprint" for a speaker—the basic model, you might say—and right there in the upper-left-hand corner of the blueprint he's got something called the Conceptualizer. The Conceptualizer figures out what the system's got to say and delegates that job to the guys down in the scene shop, who then put the words together and figure out the grammatical relations. The Conceptualizer is the boss, who sets the specs for what's going to be said. Levelt writes a whole book showing how to fit all the results into a framework in which there's this initial Conceptualizer giving the rest of the system a preverbal message. The Conceptualizer decides, "O.K., what we have to do is insult this guy. Tell this bozo that his feet are too big." That gives the job to the rest of the team, and they put the words together and out it comes: "Your feet are too big!"

The problem is, How did the Conceptualizer figure out what to tell the language system to say? The linguists finesse the whole problem. They've left the Central Meaner in there, and all they've got is somebody who translates the message from mentalese into English—not a very interesting theory. The way around this, once again, is to have one of these Pandemonium models, where there is no Central Meaner; instead, there are all these little bits of language saying, "Let me do it, let me do it!" Most of them lose, because they want to say things like "You big meanie!" and "Have you read any good books lately?" and other inappropriate things. There's this background struggle of parallel processors, and something wins. In this case, "Your feet are too big!" wins, and out it comes.

What about the person who said it? Did he mean it? Well, ask him. The person who said it will say, "Well, yeah, I meant it. I said it. I didn't correct it. My ears aren't burning. I'm not blushing. I must have meant it." He has no more access into whether he meant it in any deep, deep sense than you do. As E.M. Forster once remarked, "How do I know what I think until I see what I say?" The illusion of the Central Meaner is still there, because we listen to ourselves and we endorse what we find ourselves saying. Right now, all sorts of words are coming out of my mouth and I'm fairly happy with how it's going; every now and then I correct myself a bit, and if you ask me whether I mean what I say, sure I do—not because there's a subpart of me, a little subsystem, which is the Central Meaner, giving the marching orders to a bunch of lip-flappers. That's a terrible model for language.

Pandemonium makes a better model: Right now, all my little demons are conspiring; they've formed a coalition, and they're saying, "Yeah, yeah, basically the big guy is telling the truth!"

Since publishing *Consciousness Explained*, I've turned my attention to Darwinian thinking. If I were to give an award for the single best idea anyone has ever had, I'd have to give it to Darwin, ahead of Newton and Einstein and everyone else. It's not just a wonderful scientific idea; it's a dangerous idea. It overthrows, or at least unsettles, some of the deepest beliefs and yearnings in the human psyche. Whenever the topic of Darwin's idea comes up, the temperature rises, and people start trying to divert their own attention from the real issues, eagerly squabbling about superficial controversies. People get anxious and rush to take sides whenever they hear the approaching rumble of evolution.

A familiar diagnosis of the danger of Darwin's idea is that it pulls the rug out from under the best argument for the existence of God that any theologian or philosopher has ever devised: the Argument from Design. What *else* could account for the fantastic and ingenious design to be found in nature? It must be the work of a supremely intelligent God. Like most arguments that depend on a rhetorical question, this isn't rock-solid, by any stretch of the imagination, but it was remarkably persuasive until Darwin proposed a modest answer to the rhetorical question: natural selection. Religion has never been the same.

At least in the eyes of academics, science has won and religion has lost. Darwin's idea has banished the Book of Genesis to the limbo of quaint mythology. Sophisticated believers in God have adapted by reconceiving God as a less anthropomorphic, more abstract entity—a sort of blank, unknowable Source of Meaning and Goodness. Some unsophisticated believers have tried desperately to hold their ground by concocting creation science, which is a pathetic imitation of science, a ludicrous parade of self-delusions and pious nonsense. Stephen Jay Gould and many other scientists have rightly exposed and condemned the fallacies of creationism. Darwin's idea is triumphant, and it deserves to be.

And yet, and yet. All is not well. There are good and bad Darwinians, it seems, and nothing so outrages the authorities as the "abuse" of Darwin's idea. When the smoke screens are blown away, they can all be seen to have a common theme: the fear that if Darwin is right, there's no room left in the universe for genuine meaning. This is a mistake, but it hasn't been properly exposed yet.

When Steve Gould exhorts his fellow evolutionists to abandon "adaptationism" and "gradualism" in favor of "exaptation" and "punctuated equilibrium," the issues are clearly not just scientific but political, moral, and philosophical. Gould is working vigorously, even desperately, to protect a certain vision of Darwin's idea. But why?

Sociobiologists claim to have deduced from Darwin's theory important generalizations about human culture, and particularly about the origin and status of our most deeply held ethical principles. When Gould and others mount their attacks on the "specter of sociobiology," the issue is presented as political: scientists on the left attacking pseudoscientists on the right. The creationists are obvious pseudoscientists. The sociobiologists are more pernicious, according to Gould et al., because it's not so obvious that what they say is nonsense. There's some truth to this, but at the heart of the controversy lies something deeper. Why do these critics want so passionately to believe that sociobiology could not be good science?

Some people hate Darwin's idea, but it often seems that even we who love it want to exempt ourselves from its dominion: "Darwin's theory is true of every living thing in the universe—except us human beings, of course." Darwin himself fully realized that unless he confronted head-on the descent of man, and particularly man's mind, his account of the origin of the other species would be in jeopardy. His followers, from the outset, exhibited the same range of conflicts visible in today's controversies, and some of their most important contributions to the theory of evolution were made in spite of the philosophical and religious axes they were grinding.

I'm not purporting to advance either revolution or reform of Darwinian theory. I'm trying to explain what Darwinian theory is and why it's such an upsetting idea.

MARVIN MINSKY: Dan Dennett is our best current philosopher. He is the next Bertrand Russell. Unlike traditional philosophers, Dan is a student of neuroscience, linguistics, artificial intelligence, computer science, and psychology. He's redefining and reforming the role of the philosopher. Of course, Dan doesn't understand my society-of-mind theory, but nobody's perfect.

ROGER PENROSE: Dan Dennett is obviously somebody who'll listen to arguments. The title of his book *Consciousness Explained* is overstated, however. I certainly don't believe that those ideas explain consciousness. He's exploring what I call "point of view A," in the list of four viewpoints I discuss in *Shadows of the Mind,* which I call A, B, C, and D. A is the strong artificial-intelligence viewpoint: that is, that mentality is to be understood in terms of computation. It doesn't matter what's doing the computation; a computer or a biological structure would be equally good.

Point of view B—which is more the philosopher John Searle's viewpoint, as I understand it—is that you could simulate the action of the brain, but the simulation wouldn't have mental attributes, so there's something other than computation involved in conscious thinking. That differs from point of view C, which is my own point of view and asserts that you can't even simulate conscious activity. What's going on in conscious thinking is something you couldn't properly imitate at all by computer, according to C.

Point of view D asserts that you can't understand mentality in terms of science at all. So I'm saying, "Yes, it's science, but it's science of a kind that eludes computation." Dennett belongs to the A point of view, of which he's one of the best exponents. Another person representing this point of view is Hans Moravec, who's written an interesting book, where he takes this point of view to its extreme and argues that within thirty-five years or so the computer will achieve our level and then race beyond us.

There are at least two different kinds of arguments you can use against A. One is the John Searle type of argument, which is that just because something carries out computations, that doesn't make it capable of being aware of anything. That argument has quite a lot of power to it. But my argument is different, because it argues against both A and B. It's a stronger argument, because it says that you can't even properly simulate conscious actions. If something behaves as though it's conscious, do you say it is conscious? People argue endlessly about that. Some people would say, "Well, you've got to take the operational viewpoint; we don't know what consciousness is. How do you judge whether a person is conscious or not? Only by the way they act. You apply the same criterion to a computer or a computer-controlled robot." Other people would say, "No, you can't say it feels something merely because it behaves as though it feels something." My view is different from both those views. The robot wouldn't even behave convincingly as though it was conscious unless it really was—which I say it couldn't be, if it's entirely computationally controlled.

ROGER SCHANK: Dan Dennett is the AI person's dream philosopher. We had all those years of putting up with philosophers like Hubert Dreyfus, who felt the need to attack AI without any attempt to understand it. Dan has made a real effort to understand AI and cognitive science, and he is the consummate philosopher in our world. I always enjoy listening to him; he always says clever things; he's one of the great fun people in our field.

What philosophers are doing is trying to put into perspective things that other people have thought. Dan does more than that, of course. He has his own thoughts, too. But it's not as if there's stuff that an AI person is likely to learn from a philosopher that will help them in AI. It's interesting to read philosophy, but it doesn't give you something you could somehow put into a program.

NICHOLAS HUMPHREY: Dan's a purist, who can be tough-minded to a fault. He's wedded to the way of looking at things he learned from Gilbert Ryle, at Oxford. Its roots are in logical positivism and behaviorism. Basically it prescribes what you can talk about and what you can't: the meaning of statements lies in the way you would verify them by observation, and if you can't offer any sort of verification, forget it. Dan got trapped by the beauty of this approach. And if it meant denying the reality of things we all know are important—like sensations, raw feelings, all the qualitative aspects of consciousness—too bad. You have to be brave to be a philosopher. You have to follow where your arguments take you, until you get proved wrong. And since no one has proved Dan wrong, he's still out there.

Of course, part of Dan is uneasy about where his theories have taken him. He's much too sensitive not to be. He realizes there's something missing. When his critics point out what they think are the weaknesses, he hates it and demands they say just what they mean. Often as not, they're reduced to mumbling, because it really is very hard to fault Dan's theory on its own terms. But I suspect that if anyone is aware of the problems, Dan himself is. It's just that he's not going to surrender to people who haven't understood him to begin with. He's not going to give way to people who challenge him on wishy-washy metaphysical grounds.

Dan's book *Consciousness Explained* is tremendously original, and it's already having a huge impact on cognitive psychology. He's produced the best account yet—a brilliant, funny, beautifully written description of the inner processes underlying thought. But while it's so good on the question of thinking, it's much less good on the question of feeling.

If you're going to explain "consciousness," you have to come to grips with the kind of consciousness that really counts with ordinary people. What do people want to have explained? What do they mean by consciousness? Or rather—since they may mean different things at different times—what is it they really care about?

If you listen to the kinds of questions people ask about consciousness—"Are babies conscious?" "Will I be conscious during the operation?" "How does my consciousness compare with yours?" and so on—you find again and again that the central issue isn't thinking but feeling. What concerns people is not so much the stream of thoughts that may or may not be running through their heads as the sense they have of being alive in the first place: alive, that is, as embodied beings interacting with an external world at their own body surfaces and subject to a spectrum of sensations—pain in their feet, taste on their tongue, color at their eyes.

What matters in particular is the subjective quality of these sensations: the peculiar painfulness of a thorn, the saltiness of an anchovy, the redness of an apple, the "What it's like" for us when the stimuli from these external objects meet our bodies and we respond. Thoughts may come and thoughts may go. A person can be conscious without thinking anything. But a person simply cannot be conscious without feeling.

Here's the paradox, though. What figures so strongly in ordinary people's conception of what matters about consciousness figures hardly at all in Dan's account of it. In *Consciousness Explained*, there's hardly anything about sensory phenomenology. Once when I said that in print, Dan pointed out to me in no uncertain terms that I'd ignored the several passages in the book where he does talk about sensations. Well, O.K., it's there if you look for it. There are some

passages where he talks about sensations and feelings as complex behavioral dispositions (which is, I think, on the right lines, provided
you allow that their complexity may mean that they're qualitatively
in a different league from anything else.) But my point is that for
Dan, the question of sensory phenomenology is no more than a side
issue, never the central mystery it is for me.

FRANCISCO VARELA: While Dan focuses on the cognitive level, my own
approach is to think about all levels, perhaps because I'm influenced
by the broad idea of nonrepresentationalist knowledge. In my reality,
knowledge coevolves with the knower and not as an outside, objective representation.

Dan is against the idea of experience bearing on science. I'm not
very fond of doing psychological readings of people. I do have this
distinct impression from a long discussion with Dennett, who, unlike
Minsky, is somebody you can engage in conversation and who will
read the other person's point of view. It's a delight to have a debate
with him. For reasons I still don't understand, he has an absolute
panic of bringing experience and the subjective element into the field
of explaining consciousness.

Dennett doesn't deny that people have minds. He says those
minds can be useful only if you treat them as overt behavior, as an
anthropologist does with a foreign culture. You take them at face
value. If you tell me you're in pain, I believe you. Then I note it down
in my book. Then I consider it as overt behavior. That's what he calls
heterophenomenology—or, more classically, the intentional stance.
He treats you as if you're something capable of intentionality.

I find that far too weak to support a theory of consciousness, because it's just one leg. The other leg, which is the real phenomenology—that is, the "as is," firsthand, direct account of the quality of
experience, is irreducible. To the extent that it's irreducible, his
whole enterprise just falls short of getting down to the tack we need
to get down to. On the positive side, what Dennett has done, probably better than anybody else, in terms of theorizing and writing, is to
eliminate what he calls the ghost of the Cartesian Theater. He argues
that you have to take this distributed phenomenon of the emergent
properties of the brain in order to account for consciousness. In that
sense, it's quite brilliant.

The other thing I credit him for is that he's introduced in philosophy of mind a style of discussion that's very rich: he comes into it with a philosopher's discipline, but he takes into account results from empirical research. You can't say that of people like John Searle, who keeps talking about philosophy of mind in a very dry, abstract, armchair way. I like Dennett's pulling up his sleeves and going into the lab with people. He's done something quite revolutionary, which is to steep himself in the scientific literature.

W. DANIEL HILLIS: Dan Dennett is my favorite philosopher, because he takes the trouble to understand things. I get annoyed at the traditional school of philosophy, whose members believe that they already understand things, so they pontificate on artificial intelligence without having the slightest idea of what the work is. Dennett, although I often disagree with him, takes the trouble to read the technical literature, and understand what people are doing in areas like linguistics, artificial intelligence, biology. His philosophical ideas are informed. They are sometimes wrong, but at least they are informed wrong as opposed to uninformed wrong.

Dennett sometimes is a sucker for a reductionist theory that seems to explain something. Maybe he inherits that disease from biologists. For instance, I think he's been suckered a little bit by Richard Dawkins' view of genes as the central player, because it appears to explain things. People might even argue that he's been suckered by the simple theories of AI into believing that they explain too much about the mind. Fundamentally, he's a reductionist, and he does believe that the phenomena we see in the mind are the result of fundamental physical principles. That's a philosophical standpoint I'm basically comfortable with. It maybe makes him more popular among the scientists than among the philosophers, because if he's right then all philosophy is just a matter of science that hasn't been done yet.

Dennett's ideas are compatible with the notion of science that there's a reality out there; it's understandable; it's based on some simple underlying laws, and we just need to understand what those laws are and the connection between them and what we see. Philosophers have always felt that there's a set of things that don't fit that paradigm. People used to say, "Well, the laws may apply on Earth, but they're not true for the heavenly bodies." Then, after Galileo, they said, "Well, that might be true for physical bodies, but it's not true for biological organisms." After Darwin, they said, "Well, that might be true for our bodies, but it's not true for our minds." And so on. We are backing the philosophers into a corner and giving them less and less to talk about. In some sense, Dennett is cooperating with the enemy by helping us back the philosophers into a smaller and smaller corner, and I like that.

BRIAN GOODWIN: Dennett's concept of relational order in relation to the brain is something I find extremely interesting. He suggests that the properties of mind aren't material properties, they're relational properties. That leads to the strong AI position. I tend to take a similar view with respect to artificial life—a view similar to the strong AI position, the idea that you can actually get intelligence in systems that aren't constituted of molecules and cells. You can get life in computers.

STEVEN PINKER: I've always been interested in Dennett's work, because he's interested in the main scientific questions I deal with—namely, how the mind is engineered; how the kinds of abilities we all take for granted, like recognizing a face or using common sense, get executed by mental software. His perspective of seeing psychology as reverse engineering is one I share in my day-to-day experimental work.

In forward engineering, you start off with an idea of what your machine is supposed to do and then you go out and design the machine. Biology, including cognitive science, is a kind of reverse engineering: you start off with a machine—namely, the human being—and you have to figure out what purpose it was designed for. The main impediment in getting other scientists to understand the complexity of intelligence is that people have minds that work so well that they're apt not to be suitably impressed by what their minds are doing, in the same way that they're apt to be unaware of what's going on when they digest food.

I enjoyed, but disagree in some ways with, Dan's discussion of consciousness in *Consciousness Explained*. I like it because Dan challenges us to come up with an argument for why we should believe that there exist some kind of raw feelings, or qualia, or subjective experience. He argues that there isn't any substance to the idea: a person with what we think of as consciousness and a zombie who behaved in the same way would be indistinguishable, as far as science is concerned.

I agree that the qualitative experience is not the key to understanding intelligence from a scientist's point of view. The scientifically tractable aspect of consciousness is not the fact that there are people or animals subjectively experiencing it, but the fact that some kinds of information are mutually accessible and others are not, and that there is therefore a portion of mental information-processing that has a different status than the rest of it. That's one sense of consciousness: information that's accessible to a particular body of information-processing involved with the current environment, and which in humans can interface with the verbal apparatus. So one can study, for example, why some information—say, an overlearned skill, like operating a stick shift when you're an experienced driver—is beneath the level of consciousness, whereas other kinds of mental processing (like how to operate a stick shift the first time you're learning to drive), where you have to reason something out consciously, step by step, is different.

Many people say that Dan's book should have been called *Consciousness Explained Away* instead of *Consciousness Explained.* (People congratulate themselves for that supposedly telling witticism, not realizing that Dan used it as a chapter heading in the book!) The reason the book is in some way unsatisfying is that there's another aspect of the problem of consciousness for which no one has yet come up with a satisfactory explanation: why there's a clear intuitive difference between one organism that feels pain and another organism that acts as if it feels pain but doesn't really feel it, or one organism or system that has the experience of seeing red when a red object is in front of it, whereas another acts identically in every way but does not have the experience. Until one addresses the problem of why that's so compelling an intuition, one isn't going to have a completely satisfying account of consciousness.

I read Dan as saying that we've been misled into thinking there's a real question there. According to Dan, there isn't. That's where I disagree: I suspect there's a real question and that it's not just an error in the way we conceptualize the problem. Perhaps our minds are simply not designed to be able to formulate or grasp the answer—a suggestion of Chomsky's that I know Dan hates. But the intuition that qualia exist is real, and as yet irreducible and inexplicable. For one thing, all our intuitions about ethics crucially presuppose the distinction between a sentient being and a numb zombie. Putting a sentient being's thumb in a thumbscrew is unethical, but putting a robot's thumb in a thumbscrew is something else. And this isn't just a thought experiment; the debates over animal rights, euthanasia, and the use of anesthetics in infant surgery depend on it.

One other area in which I disagree with Dan is the explanation of human intelligence in an evolutionary context. Dan makes heavy use of Richard Dawkins' concept of the meme—an idea that replicates, mutates, and differentially spreads in the medium of brains in the same way that a gene replicates, mutates, and differentially spreads in the medium of bodies. This is Dan's main way of placing cognition in the context of evolution, rather than having it appear by magic; thoughts are created by a process analogous to the process of natural selection. But there are many other ways of explaining the emergence of human intelligence in a nonmiraculous way. I think it's much more plausible that evolution designed a brain that's a kind of computer that can generate complex ideas, in ways that need not be analogous to the operation of natural selection itself.

There's a big difference between gene selection in the design of organisms and meme selection in the design of mind and culture. For organisms, undirected variation followed by selection is the explanation, and the only explanation, for complex design. In contrast, because the brain is a complex machine that was itself designed by selection, "mutations," or ideas, are virtually always directed, and meme dynamics need not be the design source (though I agree it plays a big role in the demography of ideas: how many copies are out there). Memes such as the theory of relativity are not the cumulative product of selection of millions of random, undirected mutations of some original idea, but each brain in the chain of production added huge dollops of value to the product in a nonrandom way.

Here's another way of putting it. I think Dan thinks that the parallelism between genetics and memetics is profound: that it's the key to exorcising the hated idea that the mind came from nowhere, that it's a magical, miraculous "skyhook," hanging in midair. According to Dan, the power of the theory of selection of replicators is that it can explain organisms and culture in the same way. Maybe so (in my view, that would be an interesting coincidence), but then again, maybe not, and if not, it shouldn't matter to Dan's larger argument—

namely, that the mind is a product of evolution. Perhaps, as the anthropologist Dan Sperber argues, the formal mechanisms that explain cultural evolution are from epidemiology, not population genetics—ideas spread like contagious diseases, not like genes.

RICHARD DAWKINS: I think of Dan Dennett as a great fountain of ideas, and he's like a fireworks display for me. On every page of his you read, you constantly put ticks in the margin. I'm never quite sure why he's classified as a philosopher rather than as a scientist; he seems to me to do the same kind of thing I do in a somewhat different field, and I greatly admire the way he thinks, the way he uses metaphors to try to get his points across. And they're elegant metaphors; they really make you feel he's hit the nail on the head. My complaint about him is that his books set you thinking so hard that you have trouble turning to the next page, because you're so busy thinking about what's on the current one.

Chapter 11

NICHOLAS HUMPHREY

"The Thick Moment"

DANIEL C. DENNETT: *Nick Humphrey is a great romantic scientist, which sounds like a contradiction in terms, but it isn't. Nick's early pioneering work in recording the firing of individual neurons in live animals, in cats, helped pave the way for work by the neuroscientists David Hubel and Torsten Wiesel. They got the 1981 Nobel Prize in physiology or medicine for their work on such single-cell recordings in cats, but it was a technique that Nick had helped develop. Very typically, once he got the technique developed, he thought, "Well, I can spend the rest of my life doing this, or I can do something else. I don't see what the residual problems are." Of course, there were lots of problems, but at any rate, typical of Nick, he wanted to turn to other things as soon as he'd done that.*

• • •

NICHOLAS HUMPHREY *is a psychologist; senior research fellow at Darwin College, Cambridge; author of* Consciousness Regained *(1983),* The Inner Eye *(1986), and* A History of the Mind *(1992).*

NICHOLAS HUMPHREY: What is it like to be ourselves? How can a piece of matter which is a human being be the basis for the experience each one of us recognizes as what it's like to be us? How can a human body and a human brain also be a human mind?

I've put forward several different answers in the last few years, and I'm no longer happy with the earlier ones. I was interested in introspection, and our intuitive knowledge of inner states of mind. I developed a theory of "reflexive consciousness," and I thought, basically, that reflexive consciousness is all that counts. You either have introspective knowledge of your own states of mind or you're not conscious at all. If this were right, consciousness would be a very high-level faculty. It might be something that has evolved only in the great apes and human beings. I suggested that it has evolved specifically to enable people to read their own and other people's minds— and so to become better "natural psychologists." This idea went down well with many people. When I published *Consciousness Regained* and *The Inner Eye*, colleagues like Richard Dawkins were full of enthusiasm: "I think Humphrey's got it! At last we have an answer to the big question: how human consciousness evolved."

I thought I'd done a good job on it. But there were problems. One of the consequences of this particular view of consciousness—consciousness identified with introspection and self-reflection—was that it meant one had to exclude from the club of conscious beings a whole lot of animals, babies, and other more primitive organisms, which don't have this level of self-reflection. The more I tried to persuade myself that, say, a rabbit in pain or a baby crying for its mother can't be conscious because they don't possess the ability to introspect, the more I got dissatisfied. I couldn't sell this idea even to myself, let alone to my nonphilosophical friends.

Surely consciousness can exist at a much lower level, exist unreflected on, just as the experience of raw being: as primitive sensations of light, cold, smell, taste, touch, pain; as the is-ness, the present tense of sensory experience, which doesn't require any further analysis or introspective awareness to be there for us but is just a state of existence. Surely *that* is what it's like to be me, or what it's like to be a dog, or what it's like to be a baby. That's what it's like to *be* conscious.

I call it the "thick moment" of consciousness. What matters is that I feel myself alive now, living in the present moment. What matters is at this moment I'm aware of sounds arriving at my ears, sight at my eyes, sensations at my skin. They're defining what it's like to be me. The sensations they arouse have quality. And it's this quality that is the central fact of consciousness.

This is where philosophers stumble. This is where the mind/body problem bites. How is it that anything going on inside a human body or inside a human brain could have a quality? How could the physiological activity underlying sensation have the conscious feel it does, how could it belong to a "sensory modality"—that is, be visual or auditory or tactile or olfactory?

I found myself thinking, Enough of this stuff about higher-level thought processes and the capacity for introspection! Enough, in a way, of all the recent advances in cognitive psychology, and AI! Enough of propositional discourse, and second-order beliefs, and so on! They're interesting problems in their own right, but the problem that people ought to be addressing is the problem of sensation.

At Cambridge, I did a degree in physiology and psychology. I was fortunate enough to have Larry Weiskrantz, an American psychologist, as my Ph.D. supervisor. At the time I was starting on my Ph.D., Weiskrantz had an experiment under way on the effects of visual-cortex lesions in monkeys. He was trying to confirm and extend the findings of the Chicago-based psychologist Heinrich Kluver that destruction of the visual cortex produces almost total blindness.

This wasn't my research project, and perhaps I should have left it alone. But Weiskrantz went away to a conference, and I got to spend some time with two blind monkeys in his lab. I hung out with them day after day, sitting with them, playing with them, interacting with them, trying to work out what was going on. I wanted to know whether they were as blind as people thought they were. It became clear after a couple of days that they were not so blind at all. When I moved my hand in front of their faces, they would follow it with their eyes. And soon enough I was able to persuade them not just to look at where my hand was but to reach out and take a piece of apple from it. The fact that they couldn't do it when the lights were down made it clear they were doing it by vision. When Weiskrantz came back from the conference, I told him I'd taught his blind monkeys to see.

I began working intensively with the monkeys, and within a few months I had brought them to the stage where they could reach out and grasp any small moving object. Weiskrantz and I quickly published a paper in *Nature* titled "Vision in a Monkey without Striate Cortex." A lot of people were amazed. It went absolutely against the standard view.

I went on working with one of the monkeys—she was called Helen—for seven years. She became a friend, a pet. I'd take her for walks around the countryside. By the end of her training, she'd progressed so far that in many ways she was just like a normal monkey. She could run around a room, avoiding obstacles, searching for nuts or currants on the floor. She had 3-D vision. She could reach out and catch a passing fly. Here was a monkey with no visual cortex, missing the apparatus required to see, and yet she was, in many respects, indistinguishable from a normal seeing monkey.

Weiskrantz made a remarkable new finding. Following up on what I discovered with Helen, he began looking for a similar capacity in human patients with damage to the visual cortex—patients who were supposed to be completely blind in the affected area of the visual field. It turned out that even though they believed themselves to be blind, they were in reality quite capable of using visual information. They could "guess" where a light was located, or what shape an object was, and get it right almost every time. They had a kind of unconscious vision, without any of the usual visual sensations, without anything that told them that light was arriving at their eyes. Weiskrantz called it blindsight. It's become a celebrated phenomenon, widely discussed by philosophers and cognitive scientists.

My own work took a different direction. I'd become uneasy about working with monkeys with brain lesions. Even though it was fascinating and exciting, I didn't want to go on doing experiments that were so hurtful to the animals. I decided to change tack entirely and study monkeys' esthetic preferences. If you give a monkey a choice of environment, what will it prefer? What do they like to look at and listen to? I soon discovered that monkeys have very strong color preferences. They like the blue/green end of the spectrum; they dislike the yellow/orange/red end. Their reactions are intense, much stronger than those of human beings. A red room really upsets them. Blue calms them down.

I'd hoped to find evidence of monkeys' showing preferences for beautiful shapes and forms and sequences, and so on. But there was very little in the way of these more sophisticated preferences: monkeys show no special liking for balance or harmony in visual forms; no interest in Mondrians or Picassos. They don't like any sort of music; in fact they always prefer silence. It's not that they have poor esthetic taste; they seem to have no interesting tastes at all.

The upshot was that I ended up writing theoretical essays about esthetics in humans rather than experimental papers about esthetics in monkeys. I wrote a paper called "The Illusion of Beauty," about the evolution of the sense of beauty. There was a lot of media interest in it. It was broadcast and won the Glaxo Writer's Prize. But I have to say that it was entirely speculative, not based on any solid evidence.

During this time at Cambridge, I ran a lab and I did a lot more experimental work, in areas such as concept formation and time perception. But something changed in me. I began to realize that while I loved doing experiments, I really wanted to get on with theorizing. So I thought I might as well leave the experimenting to other people. There were lots of good experimentalists around.

I made a decision not to go on working in a lab. I gave up my position at Cambridge. I wanted to concentrate on theoretical writing. I wrote *Consciousness Regained*, which came out in 1983. I got involved in movie stuff. I became distracted. There were different horizons around.

It was in 1987 that I went to work with Dan Dennett. It was a strange move. I had been a director of research at Cambridge; at Dan's Center for Cognitive Studies, at Tufts, I was a research assistant. I didn't feel entirely good about it, but it turned out to be a wonderful time. Over the next couple of years, Dan and I both started working on new books, Dan on *Consciousness Explained* while I wrote *A History of the Mind.*

Dan and I thought alike about a lot of things when I first went to work with him. We both had much the same view about what constitutes the central problem of consciousness—the problems of intentionality, self-reflection, and all that. But after spending some time there, I found that interacting with Dan made me realize how much was missing from his picture of the mind—and from mine. Of course, Dan is a much cleverer philosopher than I am, and, as a matter of fact, knows a good deal more cognitive psychology than I do. Maybe I needed to see how my own earlier ideas looked when Dan helped me express them more rigorously, to see what was wrong. What was wrong was that they left out the phenomenology. We were left with consciousness that didn't feel like anything.

I decided that the big problem I had to work on was the nature of conscious sensation. Dan and I used to drive up to his place in Blue Hill, Maine, weekend after weekend, furiously arguing about whether there's such a thing as the sensation of red, or the feeling of pain, or the taste of cheese. Dan would say, "Look, I hear what you're saying, but I simply don't have any reference point for it. Your raw sensations, if they exist, leave nothing behind. They might as well never have occurred." I'd say, "Yes, Dan, I know, but they just are. I'm having them now. I'm living these things." For Dan, if there's nothing left after the sensation has passed—nothing in the way of a text, which says something like, "Memo to self: have just had a sensation"—then it didn't happen.

For Dan, the basic constituents of consciousness are ideas, judgments, propositions, and so on. His problem is to explain how people come to have the particular thoughts they do: how they make decisions, pull things out of memory, construct verbal reports, and so on. To caricature it, his picture of the mind is of a kind of cerebral office, with memos, faxes, and phone calls flying around and competing for the attention of the frantic office staff. The final output of all this information processing is a "conscious text," expressed in words or their equivalent—what Dan calls the heterophenomenological text, corresponding to the stream of consciousness.

But for me, now, the basic constituents of consciousness are raw feelings or sensations. And my problem is to explain how people come to experience these sensations as such: how the "activity of sensing" results in sensations having their qualitative character, immediacy, present-tenseness, sense of belonging to the self, and so forth. My picture of the mind is more that of a cerebral cinema organ, with the organist creating music to match the mood of the film being played at the surface of the body. There doesn't even have to be a final output—at least, certainly not any kind of text—since the experience of consciousness consists essentially in the ongoing activity of playing the organ.

What it boils down to is whether or not you accept an instrumentalist criterion for "meaning": whether you say that only if something is instrumental in producing something else does it have any significance or value. This approach is closely allied to positivism and behaviorism. It also ties in with the Protestant political ethic, where everything is valued only in terms of its effect on the next generation. I suppose this ethic is part of Dan's cultural background. At any rate, it colors Dan's view of the mind. The meaning and value of a mental event consists in what can be made of it later.

Although it's hard to argue against this idea, it seems to belie the reality of our experience, the immediacy and presentness of sensations. For Dan, consciousness doesn't occur until the mind has made up a story and reported back. Consciousness is the story. I say that consciousness is the immediate reaction to the stimulus at the body surface. I make conscious sensations equivalent to an action, to an act of engagement with the stimulus.

The analogy I like to use is from art history. Until the French impressionists came along, most paintings were concerned with how a situation is developing in time: where things have come from and where they're going. It took Monet to value the present moment for itself. To say, "This is Rouen cathedral as I am experiencing it now; this is what hits my face as I look at it." The clock on Rouen cathedral in his paintings doesn't even have a hand on it. There's no time dimension here, no before and after, just a now. Monet grasped this moment, and celebrated it just for what it is, producing a thick painting, full of pigment, to represent a thick moment of his subjective experience, with no antecedents and no consequences. It's the same with the thick moment of sensation, the time we live in. Stand on a street corner in New York and look at the people passing by: the amazing thing is that they're living in the present.

The focus of almost all contemporary research in AI and cognitive science is on explaining thinking rather than feeling. It's been remarkably successful in its own terms. We already have thinking machines. We'll have better ones—fourth-generation, fifth-generation thinking machines. But we're not going to say, "Wow! This is something we could never have imagined." If, however, someone could devise a feeling machine, a machine that had conscious sensations, then we'd say "Wow!" But no one's working on that problem. IBM isn't interested in feeling machines.

Even if we did set out to design feeling machines, we'd probably be unable to design machines that had conscious feelings anything like ours. The reason is that so much depends on the particular biological history, the particular route, by which we got where we are today. Our sensations have what I call skeuomorphic features—features that derive from ancestral ways of doing things and which no longer have any relevance or payoff in today's world but nonetheless supply richness and quality. Think of the analogy with architecture. Many modern buildings contain features that derive from the way Greek or Roman temples were built but have nothing to do with the buildings' functions.

Here's a concise description of my model of reality: I'm me. I'm living an embodied existence, in the thick moment of the conscious present. I'm trying to work out why.

There's a poem I like by e.e. cummings:

> since feeling is first
> who pays any attention
> to the syntax of things
> will never wholly kiss you;

DANIEL C. DENNETT: Nick Humphrey is a great romantic scientist, which sounds like a contradiction in terms, but it isn't. Nick's early pioneering work in recording the firing of individual neurons in live animals, in cats, helped pave the way for work by the neuroscientists David Hubel and Torsten Wiesel. They got the 1981 Nobel Prize in physiology or medicine for their work on such single-cell recordings in cats, but it was a technique that Nick had helped develop. Very typically, once he got the technique developed, he thought, "Well, I can spend the rest of my life doing this, or I can do something else. I don't see what the residual problems are." Of course, there were lots of problems, but at any rate, typical of Nick, he wanted to turn to other things as soon as he'd done that.

There's a little piece he wrote for the *London Observer* a few years ago which best expresses who he is and what he is. He wrote a piece on the four-hundredth birthday of Isaac Newton, in which he compared Newton to Shakespeare and drew attention to C.P. Snow, who had said that Newton was a scientific Shakespeare. Nick said that that was wrong: if Newton hadn't done what he did, somebody else would have done it sooner or later, probably quite soon. If Shakespeare hadn't done what he did, however, *nobody* would ever have done it. Newton did things God's way; Shakespeare did things Shakespeare's way.

In this simple passage, he puts his finger on a very important difference between two kinds of creativity. It's very clear that for Nick the Shakespeare style of creativity is more enticing than the Newton style, which is an unusual attitude in a scientist. The dream of proving a famous theorem isn't as enticing to him as the dream of doing something so idiosyncratic and original that people would say, "Well, only Humphrey could have done that; this is a unique and personal contribution to world culture." You find this in the arts; you don't find it as much in the sciences.

In his theory of mind, Nick tries to support his powerful intuition that my view leaves something out. He's much smarter than most of the people who have that intuition, and he realizes that the arguments for my view are pretty strong. Unless he can come up with a radically different rival, I win, in effect. He's casting about, with great ingenuity and passion, for a rival theory. He hasn't found it, but at least he knows what it would be to find a genuine rival and not just some old-fashioned reactionary retreat into old ways of thinking. He knows those old ways won't work. I'm right about those, and he admits it. He's casting about for an entirely new way of showing what I'm leaving out, by his lights. I find it much more interesting to look at his efforts than to look at the efforts of other people who disagree with me, because most of the people who disagree with me just go back to various threadbare themes that have already been dealt with—that were exploded long ago.

NILES ELDREDGE: Nick's notion of the adaptive significance of self-consciousness, self-awareness, in humans, is a nice story. It may or may not be true. The point is that you have this inner eye and you can refer to yourself as the best estimate of what is going on in the mind of somebody with whom you're sitting around the campfire. I was always attracted to that, and that's a good example of how I'm not anti-adaptationist myself. I think evolution produces things for reasons.

STEVE JONES: I have a problem with scientists who spend most of the time looking at their own navels, trying to define what it is they're supposed to be studying. It's like a game where both teams stand around arguing about what the rules are supposed to be. That's the difficulty with the "consciousness" game. Just how do you play it, and where's the goal? Is there a goal at all—or even a game? Define what a problem is in terms accessible to a layman, and you have the beginnings of a science. If you can't, you have nothing but a series of opinions.

My feeling about most people in that field is that they'd find life more interesting if they continued to do what most of them started by doing—getting their feet wet by doing experimental work.

There's a disease of middle-aged literary men called Hearty Degeneration of the Fat; when you get old, you boom about Big Issues. G.K. Chesterton was a classic example. Scientists, I guess, have a related problem—Anguished Uncertainty of the Elderly is probably a better term. All of a sudden you forget that science is the art of the answerable and you begin to speculate about things that basically lie outside science altogether.

I'm not saying Nick Humphrey does only that; certainly not. But it's something we're all in danger of doing. Nick Humphrey is going into fields I don't find interesting. The consciousness field, the meaning-of-life field—it's always left me cold.

FRANCISCO VARELA: Nick Humphrey is trying to bring phenomenological experience to bear in the science. He does it by making distinctions within his own British analytical tradition— the distinction he makes between· sensation and perception. I was impressed that somebody from his tradition was making that effort. In that sense, *The History of the Mind* is a remarkable book.

Toward the second part of the book, he claims to have some kind of an explanation for consciousness; I didn't find that convincing, or even understandable at all; it seemed to veer literally into literature. I didn't get it. But the first part of the book was very illuminating: the idea of using direct experience—what he calls sensation—to describe scientific brain processes. His idea of "the thick moment" is a fundamental piece of this relationship between experience and brain functioning.

Chapter 12

FRANCISCO VARELA

"The Emergent Self"

STUART KAUFFMAN: *Francisco Varela is amazingly inventive, freewheeling, and creative. There's a lot of depth in what he and Humberto Maturana have said. Conversely, from the point of view of a tied-down molecular biologist, this is all airy-fairy, flaky stuff. Thus there's the mixed response. That part of me that's tough-minded and critical is questioning, but the other part of me has cottoned on to the recent stuff he's doing on self-representation in immune networks. I love it.*

• • •

FRANCISCO VARELA *is a biologist; director of research at the Centre National de Recherche Scientifique, and professor of cognitive science and epistemology at the École Polytechnique, in Paris; author of* Principles of Biological Autonomy *(1979); coauthor with Humberto D. Maturana of* Autopoiesis and Cognition: The Realization of the Living *(1980) and* The Tree of Knowledge *(1987), and with Evan Thompson and Eleanor Rosch of* The Embodied Mind *(1992).*

FRANCISCO VARELA: I guess I've had only one question all my life. Why do emergent selves, virtual identities, pop up all over the place creating worlds, whether at the mind/body level, the cellular level, or the transorganism level? This phenomenon is something so productive that it doesn't cease creating entirely new realms: life, mind, and societies. Yet these emergent selves are based on processes so shifty, so ungrounded, that we have an apparent paradox between the solidity of what appears to show up and its groundlessness. That, to me, is a key and eternal question.

As a consequence, I'm interested in the nervous system, cognitive science, and immunology, because they concern the processes that can answer the question of what biological identity is. How can you have some kind of identity that simultaneously allows you to know something, allows cells to configure their own relevant world, the immune system to generate the identity of our body in its own way, and the brain to be the basis for a mind, a cognitive identity? All these mechanisms share a common theme.

I'm perhaps best known for three different kinds of work, which seem disparate to many people but to me run as a unified theme. These are my contributions in conceiving the notion of autopoiesis—self-production—for cellular organization, the enactive view of the nervous system and cognition, and a revising of current ideas about the immune system.

Regarding the subject of biological identity, the main point is that there is an explicit transition from local interactions to the emergence of the "global" property—that is, the virtual self of the cellular whole, in the case of autopoiesis. It's clear that molecules interact in very specific ways, giving rise to a unity that is the initiation of the self. There is also the transition from nonlife to life. The nervous system operates in a similar way. Neurons have specific interactions through a loop of sensory surfaces and motor surfaces. This dynamic network is the defining state of a cognitive perception domain. I claim that one could apply the same epistemology to thinking about cognitive phenomena and about the immune system and the body: an underlying circular process gives rise to an emergent coherence, and this emergent coherence is what constitutes the self at that level. In my epistemology, the virtual self is evident because it provides a surface for interaction, but it's not evident if you try to locate it. It's completely delocalized.

Organisms have to be understood as a mesh of virtual selves. I don't have one identity, I have a bricolage of various identities. I have a cellular identity, I have an immune identity, I have a cognitive identity, I have various identities that manifest in different modes of interaction. These are my various selves. I'm interested in gaining further insight into how to clarify this notion of transition from the local to the global, and how these various selves come together and apart in the evolutionary dance. In this sense, what I've studied, say, in color vision for the nervous system or in immune self-regulation are what Dan Dennett would call "intuition pumps," to explore the general pattern of the transition from local rules to emergent properties in life. We have at our disposal beautiful examples to play around with, both in terms of empirical results and in terms of mathematics and computer simulations. The immune system is one beautiful, very specific case. But it's not the entire picture.

My autopoiesis work was my first step into these domains: defining what is the minimal living organization, and conceiving of cellular-automata models for it. I did this in the early 1970s, way before the artificial-life wave hit the beach. This work was picked up by Lynn Margulis, in her research and writings on the origins of life, the evolution of cellular life, and, with James Lovelock, the Gaia hypothesis. Humberto Maturana and I invented the idea of autopoiesis in 1970. We worked together in Santiago, during the Socialist years. The idea was the result of suspecting that biological cognition in general was not to be understood as a representation of the world out there but rather as an ongoing bringing-forth of a world, through the very process of living itself.

Autopoiesis attempts to define the uniqueness of the emergence that produces life in its fundamental cellular form. It's specific to the cellular level. There's a circular or network process that engenders a paradox: a self-organizing network of biochemical reactions produces molecules, which do something specific and unique: they create a boundary, a membrane, which constrains the network that has produced the constituents of the membrane. This is a logical bootstrap, a loop: a network produces entities that create a boundary, which constrains the network that produced the boundary. This bootstrap is precisely what's unique about cells. A self-distinguishing entity exists when the bootstrap is completed. This entity has produced its own boundary. It doesn't require an external agent to notice it, or to say, "I'm here." It is, by itself, a self-distinction. It bootstraps itself out of a soup of chemistry and physics.

The idea arose, also at that time, that the local rules of autopoiesis might be simulated with cellular automata. At that time, few people had ever heard of cellular automata, an esoteric idea I picked up from John von Neumann—one that would be made popular by the artificial-life people. Cellular automata are simple units that receive inputs from immediate neighbors and communicate their internal state to the same immediate neighbors.

In order to deal with the circular nature of the autopoiesis idea, I developed some bits of mathematics of self-reference, in an attempt to make sense out of the bootstrap—the entity that produces its own boundary. The mathematics of self-reference involves creating formalisms to reflect the strange situation in which something produces A, which produces B, which produces A. That was 1974. Today, many colleagues call such ideas part of complexity theory.

The more recent wave of work in complexity illuminates my bootstrap idea, in that it's a nice way of talking about this funny, screwy logic where the snake bites its own tail and you can't discern a beginning. Forget the idea of a black box with inputs and outputs. Think in terms of loops. My early work on self-reference and autopoiesis followed from ideas developed by cyberneticists such as Warren McCulloch and Norbert Wiener, who were the first scientists to think in those terms. But early cybernetics is essentially concerned with feedback circuits, and the early cyberneticists fell short of recognizing the importance of circularity in the constitution of an identity. Their loops are still inside an input/output box. In several contemporary complex systems, the inputs and outputs are completely dependent on interactions within the system, and their richness comes from their internal connectedness. Give up the boxes, and work with the entire loopiness of the thing. For instance, it's impossible to build a nervous system that has very clear inputs and outputs.

The next area of significant work involves applying the logic of the emergent properties of circular structures to look at the nervous system. The consequence is a radical change in the received view of the brain. The nervous system is not an information-processing system, because, by definition, information-processing systems need clear inputs. The nervous system has internal, or operational, closure. The key question is how, on the basis of its ongoing internal dynamics, the brain configures or constitutes relevance from otherwise nonmeaningful interactions. You can see why I'm not really interested in the classical artificial-intelligence and information-processing metaphors of brain studies. The brain can't be understood as a computer, in any interesting sense, and I part company with the people who think that the brain does rely on symbolic representation.

The same intuitions cut across other biological fields. Deconstruct the notion that the brain is processing information and making a representation of the world. Deconstruct the militaristic notion that the immune system is about defense and looking out for invaders. Deconstruct the notion that evolution is about optimizing fitness to live in the conditions present in some kind of niche. I haven't been directly active in this last line of research, but it's of great importance for my argument. Deconstructing adaptation means deconstructing neo-Darwinism. Steve Gould, Stuart Kauffman, and Dick Lewontin, each in his own way, have spelled out this new evolutionary view. Lewontin, in particular, has much appreciated the fact that my work on the nervous system mirrors his work with evolution.

My fourth area of concentration—the most recent one—consists of using the same concepts to revise our understanding of the immune system. Just as conventional biology understood the nervous system as an information-processing system, classic immunology understands immunology in military terms—as a defense system against invaders.

I've been developing a different view of immunology—namely, that the immune system has its own closure, its own network quality. The emergent identity of this system is the identity of your body, which is not a defensive identity. This is a positive statement, not a negative one, and it changes everything in immunology. In presenting immunology in these terms, I'm creating a conceptual scaffolding. We have to go beyond an information-processing model, in which incoming information is acted upon by the system. The immune system is not spatially fixed, it's best understood as an emergent network.

I've also carried out empirical work corresponding to these intuitions. These ideas are incarnated into new experiments, and provide new results. For example, in classical immunology you were dealing with an external response system that was always watching out for invaders. If this made sense, the system would shrink to nothing if there were no invaders. Yet when mice are raised in milieus free from external challenge, their immune systems are normal!

Classical medicine remains baffled by the spectrum of diseases known as autoimmune diseases. Why? Because autoimmune disease is outside the paradigm of immunology. There's nothing to vaccinate against; there's no bacteria coming from outside. It's something that the system does to itself. AIDS is a dramatic case of the deregulation of this coherent emergent property, much like ecological dysfunctioning. People think AIDS is an infection. This is, of course, true, but not true in the sense that once the system is infected with AIDS it triggers a condition of self-destruction of the immune system. HIV triggers a deregulation, which then amplifies itself and becomes its own nightmare. Thus when you look at the urine of an AIDS-infected patient, less than 5 percent of the dead lymphocytes are HIV-infected.

This is typical of an autoimmune condition: the system eats itself up. Consequently, it's beginning to dawn on people that looking for AIDS vaccines is a complete waste of time. From my point of view, the right approach is first to understand the nature of this global regulation. One hint of how to do this is to look for ways to reconnect the system. In this regard, autoimmune diseases are seen as a deregulation, a condition that cries for more connectedness, rather than as a condition susceptible to treatment with a vaccine. For example, look at drug addiction in terms of a social disease: Drug addicts are in some sense an autoimmune disease of society, because they end up destroying segments of society. What those people need is to be given support, jobs, and family care; you reconnect them back into the society. One approach we study is to provide new, normal antibodies that help to re-create the network. We are researching more sophisticated ways of doing this, but we need to have a pointer on where to go. Vaccines are not the answer.

I'm interested in establishing empirical correlations between a long-standing interest in Buddhist practice and scientific work. Western tradition has avoided the idea of a selfless self, of a virtual self. This egolessness, or selflessness, is truly the core of Buddhism. Over the past two thousand years, the Buddhists have developed philosophical, phenomenological, and epistemological sophistication, and they have invoked this intuition in a very hands-on way. We can use these insights much like people in the Renaissance used Greek philosophy to try to understand the science of Galileo.

Buddhism is a practice, not a belief, and every Buddhist is, in some way, lay clergy—involved in the way a scientist is involved in his or her work, or in the way a writer's mind is involved in writing, present in the background, all the time. People today have the leisure and sophistication to practice what before was only practical for monks. Buddhism affects Western culture through the individuals who practice it, through people who occasionally take it up as an escape. Buddhist ideas are prevalent throughout our culture—in physics and biology, for example, the basic ideas are Buddhism in disguise.

My view of the mind has been influenced by my interest in Buddhist thought. Buddhists are specialists in understanding this notion of a virtual self, or a selfless self, from the inside, as lived experience. This is what fascinates me about that tradition. Dan Dennett, incidentally, has come to the same conclusion in his own way. But while Dan focuses on the cognitive level, my own approach is to think about several biological levels, as I have mentioned, perhaps because I'm influenced by the broad idea of nonrepresentationalist knowledge. In my reality, knowledge coevolves with the knower and not as an outside, objective representation.

I see the mind as an emergent property, and the very important and interesting consequence of this emergent property is our own sense of self. My sense of self exists because it gives me an interface with the world. I'm "me" for interactions, but my "I" doesn't substantially exist, in the sense that it can't be localized anywhere. This view, of course, resonates with the notions of the other biological selves I mentioned, but there are subtle and important differences. An emergent property, which is produced by an underlying network, is a coherent condition that allows the system in which it exists to interface at that level—that is, with other selves or identities of the same kind. You can never say, "This property is here; it's in this component." In the case of autopoiesis, you can't say that life—the condition of being self-produced—is in this molecule, or in the DNA, or in the cellular membrane, or in the protein. Life is in the configuration and in the dynamical pattern, which is what embodies it as an emergent property.

I find it fascinating to apply this same line of analysis to my own mind, in the cognitive domain. My own sense of self, "me," can be seen in the same light. I have to be relentless to hold on to my identity. These ideas help us to come to a real appreciation of what it means to have an identity—to comprehend what we think of as our own mind. My mind has the quality of "being here" so I can relate to others. For example, I interact; but when I try to grasp it, it's nowhere—it's distributed in the underlying network.

Let me add that this emergence and nonlocality has nothing to do with the current hype about quantum mechanics and the brain. That stuff is perhaps an interesting hypothesis to entertain, but it has no scientific evidence behind it. On the other hand, I'm talking about thirty years' worth of results in cognitive science. I'd go one step further and dispute the typical physicist, who believes that he or she is dealing with fundamental reality. A physicist will say that we're made of atoms. Such statements, while true, are irrelevant. The statement "You're looking at me" doesn't have the same weight as statements concerning the cellular level. There is a reality of life and death, which affects us directly and is on a different level from the abstractions. We have to abandon the enormous deadweight of the materialism of the Western tradition, and turn to a more planetary way of thinking.

STUART KAUFFMAN: Francisco Varela is amazingly inventive, freewheeling, and creative. There's a lot of depth in what he and Humberto Maturana have said. Conversely, from the point of view of a tied-down molecular biologist, this is all airy-fairy, flaky stuff. Thus there's the mixed response. That part of me that's tough-minded and critical is questioning, but the other part of me has cottoned on to the recent stuff he's doing on self-representation in immune networks. I love it.

The work Francisco is doing on the core immune network, which is representing self, and the peripheral system, which is responding to an outside world, is very intriguing. I'm not sure whether he's correct in his thesis that the immune repertoire evolved as a means of representing self, and that an evolutionary consequence was the capacity to recognize and ward off nonself. Whether or not one agrees with that sort of ontological and evolutionary argument, the work he's doing is very nice. It's imaginative, it's tied down to facts in places where it can be tied down. He is very smart, utterly charming and graceful, and his capability in any one of a large number of languages astonishes me.

I first got to know Francisco, indirectly, in 1983, when I met Humberto Maturana in India. They'd come up with their theory of autopoiesis, which was considered gobbledygook by many tough-minded scientists if they paid any attention to it at all. After listening to Humberto, I returned to my work on autocatalytic sets, which I'd begun in 1971 and then set aside. I believe that my autocatalytic-polymer-set story is the clearest instance I know of, in terms of a formally described model, of what they mean by autopoiesis.

It's likely that 99 percent of serious biologists have never heard of Francisco. This is for two reasons. First, he's not American or English, and the bulk of serious molecular biology is done in America and England, with some being done in France, Switzerland, and Germany. Francisco, after all, comes from South America. He's not from the "right" part of the world—that is, the kind of place that usually produces biologists. Second, Francisco is a good theoretical biologist, and theory in biology is in low repute. He's done detailed simulations of immune networks and neural networks that actually function—at least on the computer—so it's good solid theoretical biology. It ties in with our work at the Santa Fe Institute on emergent collective phenomena.

I'm less florid than Francisco. Although his theoretical style may appeal to some of us theoreticians, it wouldn't appeal to tough-minded colleagues, or even to more facile experimental colleagues, who wouldn't see what the next experiment is.

This is a problem that's hard to get your mind around, if you aren't trained as a biologist. Unlike physics and chemistry, which are concept-driven and theory-driven, biology is essentially experiment- and grungy-fact-driven. Organisms are complicated, ad-hoc contraptions. That's been our view since Darwin.

Organisms are ad-hoc solutions to design problems. The standard view is that there are no deep theories of the deep meaning of ad-hoc contraptions. You take the things apart and find out how they work. Most biologists adhere to that view. Notions of underlying deep principles are not an anathema to them—they're just considered foolish.

Francisco is a philosopher, in a way. He and Humberto Maturana are right about their idea of autopoiesis. But he hasn't had a large impact in the United States. The main reason he's dismissed is that he's seen just as a philosopher. Along with Francisco, I'm among those who hold that such deep principles exist, and I'm trying to find them. I have a hard time being heard by my experimental colleagues. I would expect that Francisco has almost never been heard. In the pantheon of biological scientists, he's probably unknown.

W. DANIEL HILLIS: I used to think Francisco Varela was a mystic, because I couldn't understand his ideas. As I came to know him, I began to realize that he's actually fishing for some of the same things I am. He's trying to understand how emergent properties come from simple interactive systems. It's hard to express that question without sounding like a mystic. Cisco does not help things by genuinely being a mystic on some other issues, and hanging out with the Dalai Lama, but he's trying to get at the same issue I am. I think he's on to something, with his theories of the immune system; he's trying to look at network properties—things like attractors of the system, and so on— and trying to get above the level of looking at the chemistry of the immune system. It's yet to be seen whether that approach will actually explain anything, but I'm supportive of his quest.

Cisco clearly is a symbol for Marvin Minsky—a symbol for a set of things that Minsky is angry about. It's true that you lose perfectly good AI people when they go off into philosophy and stop doing anything useful. I think Minsky is very annoyed that one of his favorite students, Terry Winograd, started out by writing perfectly good computer programs and then went off and wrote a book on hermeneutics. That bugs Minsky, because he sees philosophy as a black hole into which his students are falling. In Marvin's mind, Cisco is a symbol of that black hole.

CHRISTOPHER G. LANGTON: Varela is one of those people who has such an engaging, articulate style of talking that when you sit and listen to him, you find yourself nodding your head and going, "Yes, yes, yes, this is all great." Then once you get out of the room, and out from under his very significant personal charm, it's hard to figure out exactly what it was he said. This is one of my problems with the field of autopoiesis. The contribution it makes is that it allows you to talk about a set of phenomena known to us from biology in a different kind of language, and sometimes just changing the language can make you look at things in a new way.

Some people who come across phenomena such as self-organization for the first time through the writings of Varela and Humberto Maturana become real advocates of autopoiesis, because it's in the context of that language that they first come across those phenomena. I came across those phenomena in the world of biology, and in the language of biology and physics, and so I'm used to thinking about them in that language, and I don't see any benefit for someone like myself in mapping them over onto the language of autopoiesis. I don't think it adds anything to our understanding of phenomenology. Once one has gone through the translation, there's no value added. It's just another way of describing the same phenomena—a way that's not particularly useful to me.

Varela would claim that he is adding something to the scientific discussion when he casts all these phenomena in his language, but whatever it is he adds always seems to slip away from me whenever I try to pin it down. I was troubled when a friend of mine pointed out that he could go through one of Varela's papers and replace the phrase "autopoietic system" with the phrase "living system" and it wouldn't change anything; in fact, several of the statements simply became tautologies. In other words, autopoiesis doesn't get me anywhere I haven't already been.

I know a lot of people, especially in Europe, who are very influenced by autopoiesis, and who are very careful in the way they describe this principle. However, I've also found that many of Varela's most ardent followers are flaming vitalists, who have found in autopoiesis a way to get beyond what they consider to be the reductionist agenda. They feel that autopoiesis allows for higher-level organizing principles in a way that what they call strict reductionist science cannot. That's epistemology, not science. The question is whether or not it's good epistemology. I don't know. Many people think it's very good, and I can't blame Varela or Maturana for the abuses wrought by their followers.

DANIEL C. DENNETT: *Post hoc ergo propter hoc!* "After this, therefore because of this." Francisco Varela is a very smart man who, out of a certain generosity of spirit, thinks he gets his ideas from Buddhism. I'd like him to delete the references to Buddhist epistemology in his writings. His scientific work is very important, and so are the conclusions we can draw from the work. Buddhist thinking has nothing to do with it, and bringing it in only clouds the real issues.

There are striking parallels between Francisco's "Emergent Mind" and my "Joycean Machines." Francisco and I have a lot in common. In fact, I spent three months at CREA, in Paris, with him in 1990, and during that time I wrote much of *Consciousness Explained.* Yet though Francisco and I are friends and colleagues, I'm in one sense his worst enemy, because he's a revolutionary and I'm a reformer. He has the standard problem of any revolutionary: the establishment is—must be—nonreformable. All its thinking has to be discarded, and everything has to start from scratch.

We're talking about the same issues, but I want to hold on to a great deal of what's gone before and Francisco wants to discard it. He strains at making the traditional ways of looking at things too wrong.

NILES ELDREDGE: I was driving in a car with Francisco in Italy once. I was just starting to watch birds, partly as a hobby and partly because so much evolutionary biology has been done on birds. I said that one neat thing about birds is that you can hear their songs, and you can also see the same color spectrum they do, so you can look at the differences in their feather patterns, and these are precisely the things that birds use to sort each other out. He got very angry and very firmly and quickly corrected me, because he had been doing a lot of research on the physiology of the vision and hearing of birds. He assured me that birds can see and hear in spectra that are way beyond human capabilities. I said I knew that, but on the other hand it was a levels problem. I was more interested in the fact that we tell the difference between birds by the songs of different species and sometimes individuals, just with our own ears, and birds are indeed using that to sort each other out—to find the correct mate, and all that.

Francisco was very formal, and impatient with the somewhat sloppy level of discourse I seem to be content with. He's interested in physiology and morphology first, and then the transformation of them, in an evolutionary sense. To me, that's where everybody has always started from, and that's why I walked away from that thirty years ago, and only got back to it tangentially. I've been studying adaptation only obliquely, being concerned mostly with the context of adaptive change. I don't intersect with his mode of thought that strongly.

BRIAN GOODWIN: The first time I ever heard of Francisco Varela was when he sent me an article on autopoiesis. He was still in Chile at the time, and I looked at it and thought it was far too abstract. I was obviously in an antiabstract phase at the time, and I put it to one side and paid no more attention to it. Then I met him.

Francisco is extraordinary in terms of the clarity of his thinking and the quality of his research, because he implements his more abstract ideas in very high-quality research work. He's an exceptional combination of a precise thinker and an imaginative thinker. Since he's in theoretical biology, he's not universally known. Anyone working in immunology will be very aware of his important contributions in that context, but his main contributions are in the realm of theory.

LYNN MARGULIS: I know some of the work of Francisco Varela, but he often talks a language I don't understand at all. I don't know if it's just me, or if he is really obscurantist. His recognition of the importance of autopoiesis, which comes from collaboration with his teacher, Humberto Maturana, involves deep understanding of living systems and how chemical self-maintenance and self-formation intrinsically define life. One part of an organism cannot be privileged over another. DNA can't be more important than membranes, because without either DNA or membranes the cell does not exist. All the components of the living system make and constantly define that system. Autopoietic systems—whether cells, organisms, or communities—are run from the inside.

Autopoiesis, as a series of criteria for defining identity and existence, applies to bacteria as well as to protoctists and people. Some say autopoiesis even applies to social systems; although debatably applied to societies, autopoiesis is a helpful organizing principle. I respect Francisco's role in recognizing the fundamental difference between living systems and engineered or other nonliving systems, but I think he obscures the way he presents his views. I don't know whether the confusion is his or mine. In this regard, Francisco is a language *Wunderkind*. I always speak with him in Spanish or French if we're alone, but when others are present we revert to English. He's totally comprehensible, articulate, and far more fluent than I am in all three languages. But there's a communication difficulty at a much deeper level. Some spectators call him a phony. I disagree. My interpretation is that he has difficulty translating his concepts into their language-trapped explanations.

Chapter 13

STEVEN PINKER

"Language Is a Human Instinct"

GEORGE C. WILLIAMS: *I'm very favorably impressed with Steven Pinker. He's going to be a superstar well into the twenty-first century. What's particularly notable is his work on the evolution of our language capability, and being able to talk about this in specific terms. There are features there that have been evolving, and that we can interpret with respect to why they evolved. I remember speculating in my 1966 book about what it is that makes the human species special. There have been all sorts of suggestions: bipedalism, tool use, that sort of thing, but it struck me at the time that the one defining capability is language.*

• • •

STEVEN PINKER *is an experimental psychologist; professor in the Department of Brain and Cognitive Sciences at MIT; director of the McDonnell-Pew Center for Cognitive Neuroscience at MIT; author of* Language Learnability and Language Development *(1984),* Learnability and Cognition *(1989), and* The Language Instinct *(1994).*

STEVEN PINKER: I call language an "instinct," an admittedly quaint term for what other cognitive scientists have called a mental organ, a faculty, or a module. Language is a complex, specialized skill, which develops in the child spontaneously without conscious effort or formal instruction, is deployed without awareness of its underlying logic, is qualitatively the same in every individual, and is distinct from more general abilities to process information or behave intelligently. (One corollary is that most of the complexity in language comes from the mind of a child, not from the schools or from grammar books.) All this suggests that language is caused by dedicated circuitry that has evolved in the human brain. It then raises the question of what other aspects of the human intellect are instincts coming from specialized neural circuitry.

I'm interested in all aspects of human language. I'm an experimental psychologist who studies language for a living: how children learn language, how people put sentences together in their minds and understand sentences in conversation, where language is situated in the brain, and how it changes over history.

My work concentrates on what science has discovered about language since 1950. In answering those questions, other questions repeatedly come up. Why is the hockey team in Toronto called the Maple Leafs instead of the Maple Leaves? Why do we say, "He flied out to center field" in baseball—why has no mere mortal ever "flown out" to center field? Why do immigrants labor with lessons and tapes and homework and English classes, while their four-year-old kids learn the language so quickly that they can make fun of their parents' grammatical errors? What language would a child speak if he was raised by wolves? I also look at what we know about how language works, how children acquire it, how people use it, and how it breaks down after injury or disease of the brain.

I unify this knowledge with three key ideas. One responds to the fact that what people do know about language is often wrong. The view of language that suffuses public discourse—that people assume both in the sciences and in the humanities—is that language is a cultural artifact that was invented at a certain point in history and that gets transmitted to children by the example of role models or by explicit instruction in schools. The corollary is that now that the schools are going to pot and people get their language from rock stars and athletes, language will steadily deteriorate, and if current trends continue we're all going to be grunting like Tarzan. I argue instead that language is a human instinct.

The second idea comes from the following: If language is a mental organ, where did it come from? I believe it came from the same source as physical organs. It's an adaptation, a product of natural selection in the evolution of the human species. Depending on how you look at it, this is either an incredibly boring conclusion or a wildly controversial conclusion. On the one hand, most people, after hearing evidence that language is an innate faculty of humans, would not be surprised to learn that it comes from the same source that every other complex innate aspect of the human brain and body comes from—namely, natural selection. But two very prominent people deny this conclusion, and they aren't just any old prominent people, but Stephen Jay Gould, probably the most famous person who has written on evolution, and Noam Chomsky, the most famous person who has written on language. They've suggested that language appeared as a by-product of the laws of growth and form of the human brain, or perhaps as an accidental by-product of selection for something else, and they deny that language is an adaptation. I disagree with both of them.

The third idea comes from the question, "Why should we be so interested in the details of language in the first place?" Language is interesting because, of course, it's distinctly human, and because we all depend on it. For centuries, language has been the centerpiece of discussions of the human mind and human nature, because it's considered the most accessible part of the human mind. The reason people are likely to get exercised by technical disagreements over the proper syntax of relative clauses in Choctaw, say, is that everyone has an opinion on human nature, and lurking beneath such discussions of language is the belief that language is the aspect of science where human nature is going to be understood first.

If language is an instinct, what does it say about the rest of the mind? I think the rest of the mind is a set of instincts as well. There's no such thing as intelligence, a capacity for learning, or a general ability to imitate role models. The mind is more like a Swiss Army knife: a large set of gadgets, language being one of them, shaped by natural selection to accomplish the kinds of tasks that our ancestors faced in the Pleistocene.

Why do I call language an instinct? Why not a manifestation of an ability to acquire culture, or to use symbols? There are four kinds of evidence that have been gathered over the last century.

One of them is universality. Universality, by itself, doesn't indicate that the ability in question is innate. For all I know, VCRs and fax machines are now close to universal across human societies. But universality is a first step to establishing innateness, and it was a remarkable and unexpected discovery—early in the century, when anthropologists first started exploring societies in far-flung parts of the globe—that without exception, every human society has complex grammar.

There's no such thing as a Stone Age language. Often you'll find that the most materially primitive culture has a fantastically sophisticated, complex language. Likewise, within a society, complex grammar is universal. To appreciate this, you first have to put aside "prescriptive grammar"—the grammar of schoolmarms and copyeditors (don't split infinitives, watch how you use "hopefully," don't let your participles dangle, don't say "them books"). That has nothing to do with what I'm talking about; it's in large part conformity to a set of conventions for a standard written dialect—something that all literate people have to master, but separate from ordinary conversation. The grammar of the vernacular, in the sense of the unconscious rules that string the words together into phrases and sentences when we converse, is far more sophisticated. If you simply try to determine what kind of mental software it would take to generate the speech of a typical person in the street, or a typical four-year-old, you'll find that it's always extremely complex and has the same overall design within a society and across societies. All languages use things like nouns and verbs, subjects and objects, cases and agreement and aux-iliaries, and a vocabulary in the thousands or tens of thousands.

Those are the first two bits of evidence, the universality of lan-guage and the universality of the design of language—that is, the kinds of mental algorithms that underlie people's ability to talk. The third bit of evidence is from my own professional specialty, language development in children. We see language development proceed the same way in all the world's cultures. It's remarkably rapid, as any parent can attest. Children begin to babble in their first year of life. First words appear at about one year of age. First word combinations, things like "more milk" and "all gone doggy," happen at about eigh-teen months. Then around the age of two, there's a burst of about six months—even less for some children—in which one sees a flowering of virtually the entire grammar of English: relative clauses, passives, questions with "WH" words, and constructions so complex that the researchers in artificial intelligence haven't been able to duplicate them in computer systems that would allow us to converse with a computer in English. Nonetheless, children have mastered these con-structions before the age of three, and you have the impression at a certain point that you're having conversations with your child, whereas the child a short time before could produce no more than one or two words of baby talk.

And what the child has done is solve a remarkably difficult com-putational problem. The problem can be stated as an engineering task: design an algorithm that will take a sample of sentences and their contexts from any of the five thousand languages on the planet, and after crunching through a number of these sentences—say, a couple of hundred thousand—come out with a grammar for the lan-guage, regardless of what the language is. That is, Japanese sentences in, Japanese grammar out; Swahili sentences in, Swahili grammar out. This problem is way beyond the capability of any current artifi-cial-intelligence system. Current natural-language processing sys-tems can't even use a single language, let alone learn to use any language. Nonetheless that's what the child does in those six months, despite the lack of grammar lessons or even feedback from parents. Moreover, if you crank up the microscope on baby talk, you often find that it conforms to universal constraints that characterize lan-guage across the planet. In the kind of experiments I do in my day-to-day work, in which you get a child in a situation where he has to use some construction he hasn't been challenged with before, the child often gets it perfect on the first shot, as if he had all the pieces and just had to let them fall together.

Children also have a remarkable ability to avoid errors. Our ears do perk up when we hear things like "breaked" and "comed" and "goed." But if we were to look at the much larger set of errors that a computer would make, because the errors would be natural conclusions to jump to about the logic of the language, in most cases it never occurs to children to make that error, even though it's the first thing a logician or a cryptographer or a computer program would guess.

Language development isn't driven by general communicative utility. The child doesn't talk better and better just to get more cookies, or to get more TV, or to be allowed to play outside more often. A lot of the changes you see in children's development simply make their speech conform better to the grammar of the language they're acquiring. Here's an example. Take a verb like "to cut," "to hit," or "to put." Children go through a stage in which they make errors like "cutted," "hitted," and "putted." A child at that stage is simply making distinctions that we adults don't. If I say "On Wednesday I cut the grass," it could mean that I cut the grass every Wednesday or that I cut the grass last Wednesday, because in English the past tense and present tense of "cut" are identical. A child who says "cutted" can distinguish the two, even though in some sense he is making a grammatical error. Children outgrow that "error," and in doing so they make their language worse in terms of the ability to communicate thoughts. What's going on in the mind of the child isn't like a hill-climbing procedure, where the better you're communicating the more you stick with what you have, but an unconscious program that synchronizes the child's language with the language of the community.

There are exotic circumstances where one can show that children are injecting complexity into the language. They're not simply repeating or reproducing imperfectly what they hear, but making the language more complex. These situations are referred to as creolization. They were first documented in cases where children in plantation or slave colonies were exposed to a mishmash or "pidgin" of choppy, ungrammatical strings of words that served as a lingua franca among the adults, who had come from different language communities. The first generation of children who were exposed to a pidgin did not reproduce that pidgin but converted it into a language with a systematic grammar called a creole. There are several cases where creolization can be seen happening today. These are cases in which deaf children are either exposed to a defective version of sign language, because their parents didn't learn it properly, or, in the case of Nicaraguan sign language, because no sign language exists and the children were recently put together in schools for the first time and are inventing, in front of our eyes, a language with a systematic grammar.

The final bit of evidence is that language seems to have neurological and perhaps even genetic specificity. That is, the brain is not a meatloaf, such that the less brain you have the worse you talk and the stupider you are, but seems to be organized into subsystems. Using brain damage and genetic deficits as tools, we can see how the brain fractionates into subcomponents.

First, there are cases in which language is impaired but intelligence is intact. For example, there are forms of aphasia, caused by strokes, in which people lose the ability to speak or understand but retain the rest of their intelligence. A slightly less extreme condition is called "specific language impairment," or SLI, in which children don't develop language on schedule or in a normal way: the language appears late and the children have to struggle with it. Pronunciation improves in adulthood, with the help of lots of therapy and practice, but the victims speak slowly, hesitantly, and with many grammatical errors. They have trouble doing certain language tasks that any five-year-old can do. For example, a tester shows a picture of a man doing something for which there doesn't exist a word, like swinging a rope over his head, and says, "Here's a man who likes to 'wug.' He did the same thing yesterday. Yesterday he . . ." A five-year-old will say "wugged," even though he's never heard "wugged" before. Presumably he creates it by applying the mental equivalent of the rule of grammar: "Add 'ed' to form the past tense." If you give this task to a language-impaired victim, very often he'll say, "Well, how should I know? I've never heard the word before." Or he'll sit and think, and reason it out as if you'd given him a calculus problem to solve; the answer doesn't come naturally.

This is despite the fact that victims of SLI are, by diagnostic definition, normal in intelligence—that is, if they weren't normal in in-

telligence they wouldn't have been classified as "specifically language-impaired." They aren't deaf, and they aren't autistic or socially
abnormal. Often, in fact, they can be superior in intelligence. There
are some children with SLI who are excellent in math but who find
speaking a pain. Specific language impairment seems to run in families—something that language therapists have known for years, because they'll treat Johnny and then a few years later they'll treat
Johnny's sister and Johnny's cousin. In the last few years, large-scale
familial and twin studies have shown that SLI is highly heritable. The
crucial study—identical twins reared apart—has not been done, because only about seventy of these pairs in the whole world have been
studied, and none of them happens to have SLI.

In cases where you find a bad gene or an injured brain, and language suffers but the rest of the brain is all right, there's always the
objection that perhaps language is the most mentally demanding
thing we do. If there is any compromise in processing power, language will suffer the most, but that doesn't indicate that language is
somehow separate from the rest of cognition; it may be just quantitatively different. The clincher is what people in my field call a double
dissociation, where one sees the opposite kind of impairment; these
are syndromes in which language is intact but the rest of intelligence
suffers—a linguistic idiot savant, who can speak, and speak well, but
is retarded. There are a number of syndromes in which that can happen, including spina bifida and Williams syndrome. In those cases,
you have what therapists call chatterboxes or blatherers; a child goes
on and on in beautifully formed sentences that often have no connection to reality. This can happen in children with an IQ of 50, who
cannot tie their shoes or handle money. That's evidence for the claim
that language is a separate mental system, an instinct.

Why do I call language an adaptation? What is the alternative?
Gould and Chomsky suggest that language is a by-product. Perhaps,
as we developed a big brain in our evolutionary history, language
came automatically, the same way that when we adopted upright
posture our backs took on an S-shaped curve. Perhaps we have language for the same reason we have white bones. No one would look
for an adaptive explanation for why bones are white as opposed to
green. They're white as a side consequence of the fact that bones
were selected for rigidity; calcium is one way to make bones rigid,

and calcium is white. The whiteness is simply an epiphenomenon, an accident.

The argument from Chomsky and Gould is that maybe language was an unavoidable physical consequence of selection for something else, perhaps analytical processing, hemispheric specialization, or an enlarged brain. No one who was around when language evolved is here to tell us about it, and words don't fossilize, so the arguments have to be indirect. However, there's a standard set of criteria in biology for when to attribute something to natural selection—that is, when it may be called an adaptation—and when to look at it as a by-product, or what Gould and Lewontin call a "spandrel." Ironically, what Gould and Chomsky have *not* done is apply these standard criteria to the case of language. They've noted the logical possibility that language doesn't have to be an adaptation, but they haven't said, "Let us now pull out the test kit, apply it to language the way we apply it to any other biological system, and see what the answer is."

The test is articulated very well by George Williams and Richard Dawkins, and that test is complex adaptive design. The fundamental problem in biology is to explain biological organization: why animals are complex arrangements of matter that do unlikely but interesting things. Dawkins and Williams noted that before Darwin, complex design was recognized as the fundamental puzzle of life, even by theologians. In fact, for them, it was an argument for the existence of God. The Reverend William Paley put it best: Imagine that you're walking across a field and you come across a rock, and you ask someone, "How did the rock get there?" and they say, "Well, the rock's always been there." You'd probably accept that as about as good an explanation as you had any right to expect. But now let's say you're walking across a field and you come across a watch, and you ask, "How did the watch get there?" and someone says, "Well, it's always been there." You wouldn't accept that explanation, because a watch is an inherently improbable arrangement of matter. You can rule out the possibility that some pattern of wind and earthquakes just happened to throw together a bunch of matter that fell into the exact configuration of springs and gears and hands and dials that you find in a watch. The watch shows uncanny signs of having been designed for the purpose of telling time, which implies some intelligent creator.

Paley's argument in the nineteenth century was that any biological organ, like the eye, is much more complex than a watch. The eye has a retina, and a lens, and muscles that move it in precise convergence, an iris that closes in response to light, and many other delicate parts. Just as a watch implies a watchmaker by virtue of its complex design, an eye implies an eyemaker—namely, God. What Darwin did was not to deny that complex design was a serious problem that needed a solution but to change the solution. The brilliance of Darwin's idea, natural selection, is that it's the only physical process ever proposed that can explain the emergence of complex design. The reason you have eyes that are uncannily designed for vision is that they're at the end of a long series of replicators, such that the better the eyes worked, the more likely the design would have made it into the next generation.

One can distinguish between the eye, which all biologists agree is the product of natural selection, and features like the whiteness of bones or the S-shape of our spine, which aren't complex gadgets or seemingly engineered systems or low-probability arrangements of matter. We don't have to invent some scenario in which animals were selected by the whiteness of their bones. There, a by-product explanation rather than adaptation is perfectly plausible.

That's the test. Apply it now to language. What we've discovered in recent studies of language is that it, too, is an improbably complex biological system. It's improbable in the sense that it's found only in one species, and improbable also in the sense that most of the things you do to a brain will disrupt the ability to use language. Moreover, like a watch or an eye, it has many finely meshing parts. There is the mental dictionary, which in a typical high-school graduate contains about sixty thousand words. There are the unconscious rules of syntax, which allow us to put words together into sentences. There are the rules of morphology, which allow us to combine bits of words, like prefixes and suffixes and stems, into words. There are the rules and processes of phonology, which massage sequences of words into a pronounceable sound pattern—what we informally call an accent. There are the mechanisms of speech production, including the shape and placement of the tongue and the larynx, which seem to have been built for speech production at the expense of another biological function, like being able to breathe while you're swallowing—which other mammals can do. There's speech perception, in which the ear can decode speech at the rate of between 15 and 45 sound units per second, faster than it can decode any other kind of signal. This is almost a miracle, because at a frequency of about 20 units per second sound merges into a low-pitched buzz, so the mouth and the ear are doing a kind of multiplexing, or information compressing and unpacking. And there is the ability of a child to learn all this in a very short period of time.

These facts suggest that the anatomy of language is complex, like the anatomy of the eye. Moreover, language is quite clearly adaptive, in the sense of inherently serving the goals of reproduction. All societies use language for patently useful things like sharing technology and inventions. Language is a major means by which people share what they have learned about the local environment. Also, social relations in the human species are largely mediated by language. We rise to power, manipulate people, find mates, keep mates, win friends and influence people by language. Moreover we, and every human society, value people who are articulate and persuasive, which certainly sets up pressures for better language.

Those two lines of evidence suggest that language meets the criteria for an adaptation and a product of natural selection. We can also test the alternative—that there's some way in which language could have arisen through another route, just as whiteness comes from making bones out of calcium. Chomsky, and many anthropologists, have speculated that a big brain was sufficient to give us language. We can test that idea, because there are people with small brains. There are dwarfs, and there is normal variation within the human species, and it's certainly not the case that people with smaller brains have more trouble with language. There are some syndromes of dwarfism where the brain is not much bigger than that of a chimpanzee. Those people are retarded, but nonetheless they have language.

Brain shape is another possibility that we can rule out as the ultimate source of language. Could it be that a generally spherical brain with a certain kind of neuron packing, through complex laws of physics we don't understand, somehow gives rise to language? Again, over the range of normal variation and of pathology, there are reports of grotesquely distorted brains, usually from hydrocephalus, sometimes cases in which the brain lines the inside of the skull like the flesh of a coconut. It's possible for a person to have that condition and nonetheless develop language on schedule. One reported case was an undergraduate student at Oxford.

If we applied those criteria to any organ we weren't as fond of—and hence likely to have strong preconceptions about—as language, we'd come to the same conclusion that we do for the eye: namely, that it's a product of natural selection.

What about the rest of the mind? In this century, starting in the 1920s, there has been a pervasive, enormous intellectual movement that treated the human mind as a general-purpose learning device and attributed its complexity to the surrounding culture. There's an obvious political motivation for this idea, in that it was a reaction to some of the racist doctrines of the nineteenth century; it seems consonant with ideals of human equality and perfectibility. One can take any infant and make him or her into anything, given the right society. People who take issue with this view have often been tarred with the epithet "biological determinist"—someone who, according to the stereotype, believes that women are biologically designed for child rearing, say, or that the poor are biologically inferior. This is a specter that hovers in the background of these discussions; both in the academy and in polite intellectual discourse, the politically correct position is that the mind is a lump of wax or a blank slate.

Carl Degler, in his book on the history of Darwinism in the social sciences, traces this credo back to two sources in the academy. One is anthropology, which contributed the idea that human cultures can vary freely and without limit and that one can therefore say nothing definitive about the human species, because somewhere there will be a tribe that demonstrates the opposite. The other is psychology, which contributed the idea of the general all-purpose learning mechanism. But both ideas have now been discredited.

The impression from anthropology that humanity is a carnival where anything is possible came in part from a tourist mentality: when you come back from a trip, you remember what was different about where you went, otherwise you might as well have stayed at home. That is, many anthropologists exaggerated the degree to which the tribes they studied were exotic and strange, both to justify their profession and to raise people's consciousness about human potential. But many of their claims have turned out either to be canards, like Margaret Mead's claims about Samoa, or to miss the forest for the trees: the anthropologists spent so much time looking for differences that they didn't notice basic categories of human experience that are found in every culture, like humor, love, jealousy, and a sense of responsibility. Language is simply the most famous example of a human universal. Donald Brown, an anthropologist at UC Santa Barbara, wrote a book called *Human Universals*, in which he scoured the archives of ethnography for well-substantiated human universals. He came up with a list of about a hundred and fifty, covering every sphere of human experience. That's my interpretation of the main lessons of anthropology. The interesting discoveries aren't about this kinship system or that form of shamanry. Underneath it all—just as, in the case of language, there's a universal design Chomsky called universal grammar—there is in the rest of culture what Donald Brown calls the universal people. He characterized the human species much the way a biologist would characterize any other species.

There has also been disillusionment with the idea that came from psychology and the study of learning—including the attempt to engineer artificial intelligence—that there's a magical learning mechanism that can acquire anything. It's an idea that sounds plausible, until you start to build one.

The main discovery of cognitive science and artificial intelligence is that ordinary people are apt to be blasé about abilities that are, upon closer examination, remarkable engineering feats, like seeing in color, picking up a pencil, walking, talking, recognizing a face, and reasoning in ordinary conversation. These are fantastically complex tasks that require their own special kinds of software. When one builds a learning system, one doesn't build a system that can learn anything; one has to build a system that can learn something very special, like a system that learns large territories, a system that learns grammar, a system that learns plant and animal species, or a system that learns particular kinds of social interactions. The only way a brain could possibly work is to have this large set of learning mechanisms, tailored to specific aspects of knowledge and experience. A general-purpose learning device is like a general-purpose tool: rather than a box full of hammers, screw drivers, and saws, one would have a single tool that does everything. That possibility is inconceivable in hardware engineering and equally inconceivable in the mental-software engineering we call psychology.

If language is innate, then how much else is? Is carburetor repair innate? Is innateness a slippery slope? Of course not! The idea of a general-purpose learning device in an otherwise blank mind is so deeply entrenched that for many people it is inconceivable that there could be anything other than the two extremes: at one end, nothing is innate; on the other end, even the ability to repair carburetors is innate.

But research in psychology, linguistics, and AI have shown that there can be an interesting intermediate position. All the wonderful complex things that people do—repairing carburetors, following soap-opera plots, finding cures for diseases—might come out of the interactions among a smaller number of basic modules. The mind might have, among other things, the following: a system for intuitive mechanics—that is, our understanding of how physical objects behave, how things fall, and so forth; an intuitive biology—that is, expectations about how plants and animals work; a sense of number, the basis of mathematics and arithmetic; mental maps, the knowledge of large territories; a habitat-selection module, recognizing the kinds of environments we feel comfortable in; a sense of danger, including the emotion of fear and a set of phobias all humans have, like fear of heights and of venomous and predatory animals; intuitions about food, about contamination, about disease and spoilage and what is icky and disgusting. Monitoring of current well being: is my life going right? Is it all O.K., or should I change something? An intuitive psychology—that is, an ability to predict people's behavior from knowledge about their beliefs and desires (which, incidentally, seems to be the module that is defective in autism). A mental Rolodex, in which we store knowledge of other people and their talents and abilities. The self-concept: our knowledge of ourselves and how to package our identity for others. A sense of justice, rights, obligations. A sense of kinship, including the tendency towards nepotism. A system concerned with mating, including sexual attraction, love, and feelings of fidelity and desertion.

So, with regard to the question "Why should we care so much about language?," one answer might be that language is a human intellectual instinct, and there might be many more.

GEORGE C. WILLIAMS: I'm very favorably impressed with Steven Pinker. He's going to be a superstar well into the twenty-first century.

What's particularly notable is his work on the evolution of our language capability, and being able to talk about this in specific terms. There are features there that have been evolving, and that we can interpret with respect to why they evolved. I remember speculating in my 1966 book about what it is that makes the human species special. There have been all sorts of suggestions: bipedalism, tool use, that sort of thing, but it struck me at the time that the one defining capability is language. But nobody has ever been able to think of a reason why advanced language capability would be favored by selection. I presume that Shakespeare and Milton and Goethe did not produce an extraordinarily large number of grandchildren compared to their contemporaries of low IQ and verbal capabilities, so I speculated that maybe what evolution has tried to do is provide children with a minimal verbal capability as early as possible, so that they can have the advantage of that, and just as an incidental consequence the process develops a momentum as the individual grows, so that you end up with adults with enormously greater than required verbal capability. Pinker may be implying something of the sort. I've read some of his earlier work on the evolution of language, although not yet his book *The Language Instinct*. I will certainly do so soon.

DANIEL C. DENNETT: What I find particularly interesting about Steve Pinker is the clarity and resoluteness with which he turned his back on the ethos of MIT, where he was raised. This is somebody who was certainly educated in a very narrowly pinched and mandarin view about the nature of language and of cognitive science, and it involved giving no ground at all to evolutionary considerations. When I first met Steve, he seemed to me to be the perfect avatar of that attitude, the ultimate MIT cognitive-science product. But he's so smart; he saw the light, and shifted ground quite decisively and with great effect. That was wonderful to see.

The light he saw was evolution. What's particularly nice is that he overthrew the shackles of his education without rancor, without going overboard in the other direction. He simply saw that there was another way of looking at things, and he pursued it. I've particularly benefited from the position he developed with one of his graduate students, Paul Bloom, in a 1990 paper entitled "Natural Language and Natural Selection." They start with the standard MIT position that the language organ, as Chomsky has called it, is innate. Of

course, it's no longer really in dispute that there are aspects of linguistic competence that are innate and specific to human beings. But
then they went on to say, in a most un-Chomskyan way: Look at how
much of this innate competence can be accounted for in adaptationist terms. Look at how much of this can be explained by natural selection.

One of the motivations for resistance to the Chomskyan view was
that it seemed to be invoking magic at a crucial point. At least, the
behaviorists—who viewed language as something learned by a general-purpose learning mechanism—were clear that they wanted a no-
nonsense, no-miracle theory of how each human being comes to
have language. It's not a gift from God, it's something that has to develop, has to be designed, has to emerge from an elaborate process of
R & D, as you might say. Chomsky seemed to be saying, No, it isn't
learned, it's innate in the individual, just a God-given language organ.
That, if you stop there, is just anathema to anybody of scientific tem-
perament. It can't be that way. Pinker has driven that point home to
people.

W. DANIEL HILLIS: Growing up in the Minsky School, I was always
taught to be wary of linguists, because Minsky had a very strong reaction against the Chomsky School. I would characterize that school as
studying language without studying the fact that people are talking
about anything. That's always made me very wary of anybody who
talked about hardwiring. Steven Pinker is perhaps the first to make
me realize that linguists have something to offer, because he can talk
about his linguistic ideas from a computational viewpoint and link
them to psychological phenomena in a sensible and understandable
way. It's amazing that you put a human on earth and three years later
that human is speaking natural language. It's a phenomenon that requires a lot of explanation.

STEPHEN JAY GOULD: I don't know Steve Pinker very well. I certainly
appreciate his expositions of the Chomskyan worldview, but I sure
wish I could persuade him that adaptation is not the way to go in understanding brain function. He seems quite implacable, though.

Chapter 14

ROGER PENROSE
"Consciousness Involves Noncomputable Ingredients"

LEE SMOLIN: *Roger Penrose is the most important physicist to work in relativity theory except for Einstein. He's the most creative person and the person who has contributed the most ideas to what we do. He's one of the very few people I've met in my life who, without reservation, I call a genius. Roger is the kind of person who has something original to say— something you've never heard before—on almost any subject that comes up.*

• • •

ROGER PENROSE *is a mathematical physicist; Rouse Ball Professor of Mathematics at the University of Oxford; author of* Techniques of Differential Topology in Relativity *(1972),* Spinors and Space-time, *with W. Rindler, 2 vols. (1984, 1986),* The Emperor's New Mind: Concerning Computers, Minds, and the Laws of Physics *(1989), and* Shadows of the Mind: A Search for the Missing Science of Consciousness *(1994); coeditor with C.J. Isham and Dennis W. Sciama of* Quantum Gravity 2: A Second Oxford Symposium *(1981), and with C.J. Isham of* Quantum Concepts in Space and Time *(1986).*

ROGER PENROSE: My main technical interest is in twistor theory—a radical approach to space and time—and, in particular, how to fit it in with Einstein's general relativity. There's a major problem there, in which some progress was made a few years ago, and I feel fairly excited about it. It's ultimately aimed at finding the appropriate union between general relativity and quantum theory.

When I was first seriously thinking of getting into physics, I was thinking more in terms of quantum theory and quantum electrodynamics than of relativity. I never got very far with quantum theory at that stage, but that was what I started off trying to do in physics. My Ph.D. work had been in pure mathematics. I suppose my most quoted paper from that period was on generalized inverses of matrices, which is a mathematical thing that physicists hardly ever mention. Then there were the nonperiodic tilings, which relate to quasi crystals, and therefore to solid-state physics to some degree. Then there's general relativity. What I suppose I'm best known for in that area are the singularity theorems that I worked on along with Stephen Hawking. I knew him when he was Dennis Sciama's graduate student; I've known him for a long time now. But the main things I've done in relativity apart from that have to do with spinors and with asymptotic structure of spacetimes, relating to gravitational radiation.

I believe that general relativity will modify the structure of quantum mechanics. Whereas people usually think that in order to unite quantum theory with gravity theory you should apply quantum mechanics, unmodified, to general relativity, I believe that the rules of quantum theory must themselves be modified in order for this union to be successful.

There's a connection between this area of physics and consciousness, in my opinion, but it's a bit roundabout; the arguments are negative. I argue that we shall need to find some noncomputational physical process if we're ever to explain the effects of consciousness. But I don't see it in any existing theory. It seems to me that the only place where noncomputability can possibly enter is in what is called "quantum measurement." But we need a new theory of quantum measurement. It must be a noncomputable new theory. There is scope for this, if the new theory involves changes in the very structure of quantum theory, of the kind that could arise when it's appropriately united with general relativity. But this is something for the distant future.

Why do I believe that consciousness involves noncomputable ingredients? The reason is Gödel's theorem. I sat in on a course when I was a research student at Cambridge, given by a logician who made the point about Gödel's theorem that the very way in which you show the formal unprovability of a certain proposition also exhibits the fact that it's true. I'd vaguely heard about Gödel's theorem—that you can produce statements that you can't prove using any system of rules you've laid down ahead of time. But what was now being made clear to me was that as long as you believe in the rules you're using in the first place, then you must also believe in the truth of this proposition whose truth lies beyond those rules. This makes it clear that mathematical understanding is something you can't formulate in terms of rules. That's the view which, much later, I strongly put forward in my book *The Emperor's New Mind*.

There are possible loopholes to this use of Gödel's theorem, which people can pick on, and they often do. Most of these counterarguments are misunderstandings. Dan Dennett makes genuine points, though, and these need a little more work to see why they still don't get around the Gödel argument. Dennett's case rests on the contention that we use what are called "bottom-up" rather than "top-down" algorithms in our thinking—here, mathematical thinking.

A top-down algorithm is specific to the solution of some particular problem, and it provides a definite procedure that is known to solve that problem. A bottom-up algorithm is one that is not specific to any particular problem but is more loosely organized, so that it learns by experience and gradually improves, eventually giving a good solution to the problem at hand. Many people have the idea that bottom-up systems rather than top-down, programmed algorithmic systems are the way the brain works. I apply the Gödel argument to bottom-up systems too, in my most recent book, *Shadows of the Mind*. I make a strong case that bottom-up systems also won't get around the Gödel argument. Thus, I'm claiming, there's something in our conscious understanding that simply isn't computational; it's something different.

A lot of what the brain does you could do on a computer. I'm not saying that all the brain's action is completely different from what you do on a computer. I am claiming that the actions of consciousness are something different. I'm not saying that consciousness is beyond physics, either—although I'm saying that it's beyond the physics we know now.

The argument in my latest book is basically in two parts. The first part shows that conscious thinking, or conscious understanding, is something different from computation. I'm being as rigorous as I can about that. The second part is more exploratory and tries to find out what on earth is going on. That has two ingredients to it, basically.

My claim is that there has to be something in physics that we don't yet understand, which is very important, and which is of a noncomputational character. It's not specific to our brains; it's out there, in the physical world. But it usually plays a totally insignifi-cant role. It would have to be in the bridge between quantum and classical levels of behavior—that is, where quantum measurement comes in.

Modern physical theory is a bit strange, because one has two levels of activity. One is the quantum level, which refers to small-scale phenomena; small energy differences are what's relevant. The other level is the classical level, where you have large-scale phenomena, where the roles of classical physics—Newton, Maxwell, Einstein—operate. People tend to think that because quantum mechanics is a more modern theory than classical physics, it must be more accurate, and therefore it must explain classical physics if only you could see how. That doesn't seem to be true. You have two scales of phenomena, and you can't deduce the classical behavior from the quantum behavior any more than the other way around.

We don't have a final quantum theory. We're a long way from that. What we have is a stopgap theory. And it's incomplete in ways that affect large-scale phenomena, not just things on the tiny scale of particles.

Current physics ideas will survive as limiting behavior, in the same sense that Newtonian mechanics survives relativity. Relativity modifies Newtonian mechanics, but it doesn't really supplant it. Newtonian mechanics is still there as a limit. In the same sense, quantum theory, as we now use it, and classical physics, which includes Einstein's general theory, are limits of some theory we don't yet have. My claim is that the theory we don't yet have will contain noncomputational ingredients. It must play its role when you magnify something from a quantum level to a classical level, which is what's involved in "measurement."

The way you treat this nowadays, in standard quantum theory, is to introduce randomness. Since randomness comes in, quantum theory is called a probabilistic theory. But randomness only comes in when you go from the quantum to the classical level. If you stay down at the quantum level, there's no randomness. It's only when you magnify something up, and you do what people call "make a measurement." This consists of taking a small-scale quantum effect and magnifying it out to a level where you can see it. It's only in that process of magnification that probabilities come in. What I'm claiming is that whatever it is that's really happening in that process of magnification is different from our present understanding of physics, and it is not just random. It is noncomputational; it's something essentially different.

This idea grew from the time when I was a graduate student, and I felt that there must be something noncomputational going on in our thought processes. I've always had a scientific attitude, so I believed that you have to understand our thinking processes in terms of science in some way. It doesn't have to be a science that we understand now. There doesn't seem to be any place for conscious phenomena in the science that we understand today. On the other hand, people nowadays often seem to believe that if you can't put something on a computer, it's not science.

I suppose this is because so much of science is done that way these days; you simulate physical activity computationally. People don't realize that something can be noncomputational and yet perfectly scientific, perfectly mathematically describable. The fact that I'm coming into all this from a mathematical background makes it easier for me to appreciate that there are things that aren't computational but are perfectly good mathematics.

When I say "noncomputational" I don't mean random. Nor do I mean incomprehensible. There are very clear-cut things that are noncomputational and are known in mathematics. The most famous example is Hilbert's tenth problem, which has to do with solving algebraic equations in integers. You're given a family of algebraic equations and you're asked, "Can you solve them in whole numbers? That is, do the equations have integer solutions?" That question— yes or no, for any particular example—is not one a computer could answer in any finite amount of time. There's a famous theorem, due to Yuri Matiyasevich, which proves that there's no computational way of answering this question in general. In particular cases, you might be able to give an answer by means of some algorithmic procedure. However, given any such algorithmic procedure, which you know doesn't give you wrong answers, you can always come up with an algebraic equation that will defeat that procedure but where you know that the equation actually has no integer solutions.

Whatever understandings are available to human beings, there are—in relation particularly to Hilbert's tenth problem—things that can't be encapsulated in computational form. You could imagine a toy universe that evolved in some way according to Hilbert's tenth problem. This evolution could be completely deterministic yet not computable. In this toy model, the future would be mathematically fixed; however, a computer could not tell you what this future is. I'm not saying that this is the way the laws of physics work at some level. But the example shows you that there's an issue. I'm sure the real universe is much more subtle than that.

The Emperor's New Mind served more than one purpose. Partly I was trying to get a scientific idea across, which was that noncomputability is a feature of our conscious thinking, and that this is a perfectly reasonable scientific point of view. But the other part of it was educational, in a sense. I was trying to explain what modern physics and modern mathematics is like.

Thus, I had two quite different motivations in writing the book. One was to put a philosophical point of view across, and the other was that I felt I wanted to explain scientific things. For quite a long time, I'd felt that I did want to write a book at a semipopular level to explain certain ideas that excited me—ideas that weren't particularly unconventional—about what science is like. I had it in the back of my mind that someday I would do such a thing.

It wasn't until I saw a BBC "Horizon" program, in which Marvin Minsky and various people were making some rather extreme and outrageous statements, that I was finally moved to write the book. I felt that there was a point of view which was essentially the one I believe in, but which I had never seen expressed anywhere and which needed to be put forward. I knew that this was what I should do. I would write this book explaining a lot of things in science, but this viewpoint would give it a focus. Also it had to be a book, because it's cross-disciplinary and not something you could express very well in any particular journal.

I suppose what I was doing in that book was philosophy, but somebody complained that I hardly referred to a single philosopher—which I think is true. That's because the questions that interest philosophers tend to be rather different from those that interest scientists; philosophers tend to get involved in their own internal arguments.

When I argue that the action of the conscious brain is noncomputational, I'm not talking about quantum computers. Quantum computers are perfectly well-defined concepts, which don't involve any change in physics; they don't even perform noncomputational actions. Just by themselves, they don't explain what's going on in the conscious actions of the brain. Dan Dennett thinks of a quantum computer as a skyhook, his term for a miracle. However, it's a perfectly sensible thing. Nevertheless, I don't think it can explain the way the brain works. That's another misunderstanding of my views. But there could be some element of quantum computation in brain action. Perhaps I could say something about that.

One of the essential features of the quantum level of activity is that you have to consider the coexistence of various different alternative events. This is fundamental to quantum mechanics. If X can happen, and if Y can happen, then any combination of X and Y, weighted with complex coefficients, can also occur. According to quantum mechanics, a particle can have states in which it occupies several positions at once. When you treat a system according to quantum mechanics, you have to allow for these so-called superpositions of alternatives.

The idea of a quantum computer, as it's been put forward by David Deutsch, Richard Feynman, and various other people, is that the computations are the things that are superposed. Rather than your computer doing one computation, it does a lot of them all at once. This may be, under certain circumstances, very efficient. The problem comes at the end, when you have to get one piece of information out of the superposition of all those different computations. It's extremely difficult to have a system that does this usefully.

It's pretty radical to say that the brain works this way. My present view is that the brain isn't exactly a quantum computer. Quantum actions are important in the way the brain works, but the brain's noncomputational actions occur at the bridge from the quantum to the classical level, and that bridge is beyond our present understanding of quantum mechanics.

The most promising place by far to look for this quantum-classical borderline action is in recent work on microtubules by Stuart Hameroff and his colleagues at the University of Arizona. Eukaryotic cells have something called a cytoskeleton, and parts of the cytoskeleton consist of these microtubules. In particular, microtubules inhabit neurons in the brain. They also control one-celled animals, such as parameciums and amoebas, which don't have any neurons. These animals can swim around and do very complicated things. They apparently learn by experience, but they're not controlled by nervous systems; they're controlled by another kind of structure, which is probably the cytoskeleton and its system of microtubules.

Microtubules are long little tubes, a few nanometers in diameter. In the case of the microtubules lying within neurons, they very likely extend a good deal of the length of the axons and the dendrites. You find them from one end of the axons and dendrites to the other. They seem to be responsible for controlling the strengths of the connections between different neurons. Although at any one moment the activity of neurons could resemble that of a computer, this computer would be subject to continual change in the way it's "wired up," under the control of a deeper level of structure. This deeper level is very probably the system of microtubules within neurons.

Their action has a lot to do with the transport of neurotransmitter chemicals along axons, and the growth of dendrites. The neurotransmitter molecules are transported along the microtubules, and these molecules are critical for the behavior of the synapses. The strength of the synapse can be changed by the action of the microtubules. What interests me about the microtubules is that they're tubes, and according to Hameroff and his colleagues there's a computational action going along on the tubes themselves, on the outside.

A protein substance called tubulin forms interpenetrating spiral arrangements constituting the tubes. Each tubulin molecule can have two states of electric polarization. As with an electronic computer, we can label these states with a 1 and a 0. These produce various patterns along the microtubules, and they can go along the tubes in some form of computational action. I find this idea very intriguing.

By itself, a microtubule would just be a computer, but at a deeper level than neurons. You still have computational action, but it's far beyond what people are considering now. There are enormously more of these tubulins than there are neurons. What also interests me is that within the microtubules you have a plausible place for a quantum-oscillation activity that's isolated from the outside. The problem with trying to use quantum mechanics in the action of the brain is that if it were a matter of quantum nerve signals, these nerve signals would disturb the rest of the material in the brain, to the extent that the quantum coherence would get lost very quickly. You couldn't even *attempt* to build a quantum computer out of ordinary nerve signals, because they're just too big and in an environment that's too disorganized. Ordinary nerve signals have to be treated classically. But if you go down to the level of the microtubules, then there's an extremely good chance that you can get quantum-level activity inside them.

For my picture, I need this quantum-level activity in the microtubules; the activity has to be a large-scale thing that goes not just from one microtubule to the next but from one nerve cell to the next, across large areas of the brain. We need some kind of coherent activity of a quantum nature which is weakly coupled to the computational activity that Hameroff argues is taking place along the microtubules.

There are various avenues of attack. One is directly on the physics, on quantum theory, and there are certain experiments that people are beginning to perform, and various schemes for a modification of quantum mechanics. I don't think the experiments are sensitive enough yet to test many of these specific ideas. One could imagine experiments that might test these things, but they'd be very hard to perform.

On the biological side, one would have to think of good experiments to perform on microtubules, to see whether there's any chance that they do support any of these large-scale quantum coherent effects. When I say "quantum coherent effects," I mean things a bit like superconductivity or superfluidity, where you have quantum systems on a large scale.

NICHOLAS HUMPHREY: Roger Penrose gets full marks for effort. It was a good try. He thinks brains are capable of leaps of intuition which are not conceivably possible for a machine. He thinks human minds can see the truth or falsity of statements that are in principle noncomputable. I'm not impressed by his examples. Of course, people can do very clever and creative things that we can't yet begin to understand—nobody has a clue how Shakespeare could write his plays or Picasso paint his paintings or Hawking do his mathematics—but I don't think there's any real parallel between these astonishing achievements and noncomputable "Gödel sentences."

Penrose has got an interesting theory, but it's a theory in search of something to apply it to. I just don't think we need quite such a radical new theory to explain human intelligence and creativity.

STEVE JONES: Penrose has a strange historical tie with the Galton Laboratory, because his father was my predecessor as the head of the department. He's spoken of by mathematicians in extremely positive terms, and I'm more than willing to take that on board. I like the patterns his tiles make.

Tiling is about how you can fill a space. Seems like an obvious question: How do you tile a bathroom floor? The obvious way is with square tiles. There's another way, with diamonds. But how many other ways are there? What happens is that as you begin to go up, you get tiles of the most unexpected shapes, and you can produce tiles none of which have the same shape but in the end make a completely consistent mathematical pattern, which fills that space in both a scientifically and esthetically satisfying fashion. There's a funny bit in Francis Crick's autobiography, *What Mad Pursuit*, when he talks about visiting the Galton in the early 1950s and finding Penrose and his father playing with odd cutouts made of wood, in the hope of working out the way DNA replicated. He thought it was a complete waste of time—and it was, as far as DNA was concerned; but it was the beginning of a new branch of mathematics.

STEVEN PINKER: In *The Emperor's New Mind,* Penrose expresses some skepticism that evolution could have constructed the human mind—and is admirably clear that this aside comes more from a personal intuition than from an argument he'd be prepared to defend. It's not uncommon among some kinds of scientist to be skeptical of Darwin and natural selection. For many physicists and mathematicians, natural selection seems a repugnant kind of explanation, because it's too kludgey. It's random stochastic variation, and selection by utility seems like an ugly way to arrive at something beautiful, and for a physicist or a mathematician, or someone like Noam Chomsky, whose work has often been mathematical, the favored kind of theory is one where a conclusion can be deduced from a bunch of premises in an elegant deductive system. By the esthetic of a grammarian, or the esthetic of a physicist, natural selection seems too ugly and weak.

FRANCISCO VARELA: Roger Penrose is the perfect example of physicists acquiring an authority to speak on just about everything and anything. Between Turing, as the ideal of computation, and quantum mechanics there's something missing—a body. For Penrose, the body has disappeared. I find it amazing that because he is a famous physicist and mathematician, and probably very rightly so, he can come up with this stuff. I would say there are no clothes on Penrose.

There is an arrogance that comes with being a physicist—particularly a mathematical physicist—which also shows up in some of the crowd at the Santa Fe Institute, including Gell-Mann. Biologists, and the public at large, share a kind of physics envy.

If I have a chance to have a discussion with Penrose, I'll press him to give me just a shred of evidence that quantum processes are relevant to describing the brain. There is none. This is the same thing that happens, say, with the psychokinesis people, or the UFO people. There are shreds of things here and there, but nothing you can put on the table and bite into.

On the other hand, there are huge amounts of evidence from neurobiology and neuropsychology to make the body a very interesting set of possible interpretations which need not be computational. Penrose discovered that the mind is not computational. I agree. Then he makes this funny leap. He says, "Then it must be quantum." That's where he loses me.

W. DANIEL HILLIS: It's annoying that you get somebody who's good at mathematics who uses his mathematical credibility to pontificate on something he's speculating about. Penrose tells a good story, but he tells a fundamentally wrong story. Penrose has committed the classical mistake of putting humans at the center of the universe. His argument is essentially that he can't imagine how the mind could be as complicated as it is without having some magic elixir brought in from some new principle of physics, so therefore it must involve that. It's a failure of Penrose's imagination.

He takes a perfectly good computational idea—the idea of uncomputability—and somehow confounds that to complex behavior in humans that he can't explain. It's true that there are unexplainable, uncomputable things, but there's no reason whatsoever to believe that the complex behavior we see in humans is in any way related to uncomputable, unexplainable things. The intelligent behavior in humans is unexplainable because it's very complicated. Penrose's argument is a little bit like the arguments that the vitalists used to make about life: that life clearly couldn't be just chemistry, so therefore there must be some vital principle. Essentially Penrose is saying the same thing about the mind: that the connection between neurons firing and intelligent behavior—thinking—must involve something beyond our current understanding. He can't make that connection, therefore he thinks there must be some vital principle that has to be added. That's all there is to his argument.

RICHARD DAWKINS: Roger Penrose clearly has a massive intellect. I don't understand quantum theory enough to criticize his theory of mind, but I have gut misgivings about theories of that kind.

DANIEL C. DENNETT: Roger Penrose...I'm so glad he exists, because, as someone once said of Voltaire, if he hadn't existed, God would have had to invent him. Much the same is true of Penrose; he lucidly plays a role that needs playing, just so everyone can see it's dead wrong.

When Roger confronted the field of artificial intelligence, he tells us, he had a deep and passionate negative reaction. "Somehow," he thought, "I've got to prove that this is wrong." One way of reading what he's done is as a backhanded compliment to AI, in that what he's seen—and none of the other critics of AI have seen—is that the only way you're ever going to show that the idea of strong artificial intelligence is wrong is by overthrowing all of physics and most of biology! You're going to have to deny natural selection, and you're going to have to have a revolution in physics. The fact is that artificial intelligence is a very conservative extrapolation from what we know in the rest of science, and Penrose makes this clearer than anyone has ever done before. There's absolutely no question that he'd like nothing better than to have an absolute knock-down-drag-out refutation of artificial intelligence, and he's such an honest man—and he knows so much—that he realizes he's not going to be able to do this unless he can overthrow physics. Of course, he might be right. But he knows as well as everybody else that he doesn't have a theory yet.

Is Roger's quantum computer a skyhook or a crane? A crane is nonmiraculous; it just obeys good old mechanistic principles. A skyhook is something pretty darn special; it's either a miracle or something that requires a revolution in physics. I see Penrose trying desperately but ingeniously to invent a skyhook. He says that the brain is a sort of machine, but that you shouldn't call it a machine, because it involves quantum effects. Most biologists think that quantum effects all just cancel out in the brain, that there's no reason to think they're harnessed in any way. Of course they're there; quantum effects are there in your car, your watch, and your computer. But most things—most macroscopic objects—are, as it were, oblivious to quantum effects. They don't amplify them; they don't hinge on them. Roger thinks that the brain somehow exploits these quantum effects, so that they aren't just quantum effects going on in the background.

Two questions: First, why does he think this? Does he think there's empirical evidence that the brain is a quantum computer, and if so, what field does this evidence come from? My understanding is that he thinks that the evidence that the brain is a quantum computer comes from mathematics and nowhere else. He's now searching, trying to get assistance in this from people like Stuart Hameroff, at the University of Arizona, who argues that in the microtubules of the neurons we've got amplifiers of quantum effects.

Why? Physics certainly permits it. If you were looking for a place for a type of quantum amplifier—a little transducer of quantum effects of the brain—the microtubules would be a pretty good place. Let me just give him that, all right? Let's give Roger the claim that Hameroff has identified the site of transduction, or amplification, of quantum effects. The second question is, "What good does it do?" What architecture does Penrose have that could use these effects, that could parlay them, or exploit them, into a quantum computer of some sort? That's a tall order. Boy, if he can do that, we'll have something to look at.

ALAN GUTH: Roger Penrose is known mostly for his work in the classical theory of general relativity. Penrose is a relativity physicist. There aren't that many. There are none at MIT, none at Harvard; there's Robert Wald at the University of Chicago, Kip Thorne at Caltech, Penrose at Oxford, Rovelli at the University of Pittsburgh, Ashtekar and Smolin at Penn State. Stephen Hawking is also a classical general relativist, although his more recent work has touched other areas as well. Hawking became famous in classical general relativity, just like Penrose.

The two of them established many of the fundamental theorems that we know about the general behavior of Einstein's equations. The problems there are mainly mathematical; the theory they're dealing with is Einstein's theory of general relativity, which has been unchanged at the fundamental level since Einstein first invented it, in 1916. But nonetheless, the equations of general relativity are very complicated, and the implications of general relativity are not easy to extract.

One question, for example, that Penrose and Hawking were concerned with is what happens when matter collapses under the force of gravity to very high densities. By the mid-1960s, it was already known that matter could collapse to form a black hole, a conglomeration of mass that produces such a strong gravitational field that even light can't escape it. Nonetheless, the solutions that give rise to black holes were very special—that is, the equations could be solved only by making special assumptions about the symmetry of the collapsing matter. If the matter was perfectly spherical, you could calculate exactly how it would collapse, and you could show that it would form a black hole. If the matter was in some complicated arrangement, nobody knew for sure if a black hole would form.

Since you'd never expect the matter in the real universe to find itself in a perfectly spherically symmetric distribution, this question was very important. It was Hawking and Penrose who developed the theorems by which you can prove, without actually solving the equations for nonspherical collapse, that under certain conditions a black hole will necessarily form.

Penrose is mostly known for his work on that kind of a problem. The same kind of theorems apply to closely related questions of the initial singularity of the universe. In the standard big-bang model, one assumes that the universe is perfectly symmetrical, completely homogeneous, and completely uniform in its mass density. The real universe, of course, isn't so ideal. You make these idealizations in order to obtain equations simple enough to solve. When you run those idealized equations backward in time, you find what's called a singularity—an instant at which the mass density and temperature of the universe is literally infinite.

This is called the initial singularity. Again, there's the question of what would happen if you complicated the equations by putting in the real complexities of the real universe, with the nonuniformities in mass clumped into galaxies, which are clumped into clusters. Once you do that, the equations become clearly too complicated to solve, so what instead one needs to do is to prove general theorems about how these equations have to behave, independent of the details. That's the forte of Penrose and Hawking, and there again they were able to prove theorems that guarantee that if the universe looks anything like the universe we see, you'll find a singularity if you follow it backward in time.

LEE SMOLIN: Roger Penrose is the most important physicist to work in relativity theory except for Einstein. He's the most creative person and the person who has contributed the most ideas to what we do. He's one of the very few people I've met in my life who, without reservation, I call a genius. Roger is the kind of person who has something original to say—something you've never heard before—on almost any subject that comes up.

Part of Roger's interest in relativity from the very beginning has been a skepticism about quantum mechanics. Indeed, before he was working on general relativity he was trying to understand quantum mechanics; he was thinking about ideas like hidden variables, he was thinking about Bell's theorem and the Einstein-Podolsky-Rosen paradox. His first ideas in physics came out of applying ideas he was using to try to prove the four-color theorem, to try to understand those ideas, and only when he met the American theoretical physicist David Finkelstein did he begin to become interested in general relativity.

David Finkelstein went to London and gave a talk on his ideas about how the topology of spacetime might be different inside black holes. David was one of the very few people to think about applying topological ideas to space and time. Topology is the science of relationships without regard to actual measures, like distance; it's the study of relationships and connectivity, taken purely. Roger was a topologist; his Ph.D. was in mathematics and algebraic topology.

David Finkelstein was applying topology to the geometry of space and time. Roger, in what he was calling spin networks, was trying to build space and time up from little discrete pieces that were purely quantum mechanical. It's always been his idea—an idea shared by many others—that space and time aren't continuous; that the continuum is an illusion that has to do with the fact that we're looking at things on a large scale.

Roger had begun trying to make models of how the geometry of space might derive from little atoms of geometry, and he called these models "spin networks." They're a very deep mathematical construction, which people have recently been studying very carefully. He listened to David's talk, and told him about spin networks, and in some sense they switched places. David went home and began trying to make models of space and time as discrete processes, which is what he's been doing ever since. Roger began to think about how to apply topological ideas to the geometry of space and time.

Having invented the discrete models of space called spin networks, Roger couldn't get them to make models of space and time and incorporate relativity theory. The attempt to do so, which he's been working on since the early 1960s, is called twistor theory, and part of the reason for Roger's isolation from the mainstream of particle physics is this preoccupation with twistor theory, his efforts to formulate a complete new theory of physics that would bring together quantum mechanics and relativity theory in a new way.

Twistor theory may be concisely defined as follows: In looking at the world, we think of points—that is, things that exist in space—as being fundamental and time as something that happens to them. The fundamental thing is the things that exist, and the secondary thing is the processes through which they change in time. In twistor theory, the fundamental things in the world are the processes. The secondary things are the things that exist. They exist only by virtue of the meetings of the intersections of processes. In the twistor description of space and time, the fundamental entities are not events in space and time but processes, and the idea of twistor theory is to formulate the laws of physics in this space of processes and not in space and time. Space and time as we think about them emerge only at a secondary level.

Twistor theory is a beautiful mathematical thing. Roger, and a succession of his students, have devoted an enormous amount of effort to trying to make a fundamental theory of physics based on it. It's a deep and difficult problem; whether it's right or not is impossible to tell, and even though Roger is a genius, the work is still unfinished. We don't yet know the potential of twistor theory. Certainly it's something that only somebody like Roger could have created.

From the beginning, Roger has been very skeptical about quantum mechanics, and has always believed that quantum mechanics would not, in the end, be the correct theory, and that there was some more fundamental theory that unified quantum mechanics and spacetime. This sets him apart from many other people, who believe that quantum mechanics is essentially correct and what we need is a new dynamical theory of the geometry of space and time—in other words, that general relativity needs to be modified to something like supersymmetric gravity or string theory. Roger believes that gravity is important for understanding the puzzles of quantum mechanics, and that quantum mechanics must be modified to make room for the effects of gravity, rather than the reverse.

All Roger's thoughts are connected. The technical ideas he's thinking about in twistor theory, his philosophical thinking, his ideas about quantum mechanics, his ideas about the brain and the mind—all of them are connected.

You could say about Roger that in spite of the fact that he's the most influential living person in relativity theory, what he has accomplished is a small shadow, a faint shadow, of what his ambition has been, and continues to be.

MURRAY GELL-MANN: I don't really know Roger Penrose. I think he and I were once at Imperial College, London, at the same time, so I must have met him, but I don't remember what he looks like. I understand he has had a distinguished career in certain kinds of mathematical physics, especially the physics of general-relativistic gravitation. But recently he has put forward in a couple of popular books some ideas that I find extremely odd.

I regard self-awareness—consciousness—as being a property, like intelligence, that may eventually evolve in complex adaptive systems when they reach certain levels of complexity. I imagine that complex adaptive systems have evolved both intelligence and consciousness on enormous numbers of planets in the universe. In fact, our human levels of intelligence and self-awareness, of which we are so proud, may not be very impressive on a cosmic scale, even though they are significantly higher than those of the other apes here on Earth. Whereas I don't think it impossible in principle that we humans may someday produce computers with a reasonable degree of self-awareness. Penrose seems to attribute some special quality to self-awareness that makes it unlikely to emerge from the ordinary laws of science. He's certainly not unique in that respect; some other authors seem to react that way to the challenge of understanding consciousness. But what characterizes *his* proposal, as far as I can tell, is the notion that consciousness is somehow connected with quantum gravity—that is to say, the incorporation of Einsteinian general-relativistic gravitation into quantum field theory. I can see absolutely no reason for imagining such a thing. Moreover, we now have, in superstring theory, a brilliant candidate for a unified theory of all the elementary particles, including the graviton, along with their interactions. The theory leads, in a suitable approximation, to Einstein's general-relativistic theory of gravitation and incorporates that theory beautifully into quantum field theory in a way that avoids all the terrible problems of infinities, which plagued previous attempts to treat general relativity in quantum mechanics. We will find out someday whether superstring theory is supported by observation, for instance in experiments done with a new high-energy accelerator. But I see no basis for engaging in mystical speculation about quantum gravity.

Penrose also revives, for some reason, the long discredited idea that Gödel's work in mathematics somehow implies a special difficulty in achieving self-awareness in a physical system. I hope that Penrose will come around eventually to the simple idea that self-awareness and intelligence emerge from biology, just as biology emerges from physics and chemistry. After all, we now understand that nuclear forces arise from quark-gluon interactions and interatomic forces from electromagnetism. Hardly anyone is left who thinks that special vital forces, apart from physics and chemistry, are needed to explain biology. Well, the idea that special physical processes are needed to explain self-awareness will soon die out as well.

MARVIN MINSKY: In effect, it seems to me, Penrose simply assumes from the start precisely what he purports to prove. He asserts that humans can do certain things that we've proved, mathematically, that computers cannot do. Specifically, he suggests that humans can "intuitively" solve certain machine-unsolvable problems (such as Alan Turing's halting problem for Turing machines, or Kurt Gödel's problem of recognizing the consistency of arbitrary sets of axioms). The trouble, though, is that these problems are unsolvable only in the sense that there's no computer program that can do this and *never be wrong,* and there's absolutely no evidence that there can't be computer programs as good at intuiting—that is, guessing—as well as human mathematicians can. There's no reason to assume, as Penrose seems to do, that either human minds or computing machines need to be perfectly and flawlessly logical; as the child psychologist Jean Piaget showed, logical reasoning is a sophisticated skill that develops quite late, if at all, in normal human development. Perhaps it did not occur to Penrose that it's easy to write computer programs that can work with inconsistent sets of axioms by applying occasionally defective logic to them. Thus Penrose's assertion that no computer could ever think in a humanoid way is just that—an unsupported assertion. Where there's smoke, a sharp reader could only conclude, there is smoke.

ROGER SCHANK: Roger Penrose wrote an outrageous book on AI. It's very sad that people write books about subjects they don't understand. If you're a famous physicist, you think you have the right to comment on things that you actually don't get. There's a famous attack on AI that many people use from Gödel's theorem. It has to do with how many computations you could make in a certain amount of time. The "proof" is about how too many computations need to be made to solve certain problems in certain ways for a machine to be able to do what is necessary to think. The mistake they make is in assuming that the kinds of computations they are talking about are the kind that compose thinking. Those are probably not right assumptions; in fact everything we have learned about human thinking says they are quite wrong assumptions. The premises for these attacks usually show the ignorance of the attackers about what intelligence is all about. This includes Penrose, who really says nothing particularly interesting about AI.

Part Three

QUESTIONS OF ORIGINS

In a manner of speaking, the physicists came to the wrong book. It is interesting to note that for the most part they have little to say about the other scientists in this book, and, similarly, the other people do not comment on their work. This may have to do with the fact that the language of physics is mathematics; it may also be that ideas about complexity and evolution have not had the same relevance for cosmology and physics as they have for biology and computer science. Astronomers have studied the spectra of light emitted by distant stars billions of years ago, and have so far found no indication that the laws of physics have changed over this epoch.

Cosmology, which came into its own as a science only about thirty years ago, is concerned in part with pinning down the parameters of the universe: its expansion rate, the amount of its mass, the nature of its "dark matter." Cosmologists today are also speculating on more far-reaching questions, such as how the universe was created and how its structure was determined. While some cosmologists are speculating that the laws of physics might explain the origin of the universe, the origin of the laws themselves is a problem so unfathomable that it is rarely discussed. Might the principles of adaptive complexity be at work? Is there a way in which the universe may have organized itself? Does the "anthropic principle"—the notion that the existence of intelligent observers like us is in some sense a factor in the universe's existence—have any useful part to play in cosmology?

Particle physics, on the other hand, is a field that has been suffering from its own success. Important discoveries in the 1960s and 1970s have led to the development of the so-called standard model, a theory that appears to be consistent with every reliable particle-physics experiment that has so far been performed. The theory has too many unexplained parameters, however, to be accepted as the ultimate theory of nature, and furthermore it does not provide a quantum description of gravity. The attempts to go beyond the standard model have led particle physicists to search for a unified theory of all the elementary particles and all the forces of nature. The goal of such a theory is to unify the four fundamental forces of nature—electromagnetism, the strong and the weak nuclear force, and gravity—into an all-encompassing theory, a reductionist enterprise beyond which, it is thought, we need not go. (The unification of the first three into one or another "grand unified theory" is in sight; gravity is more of a problem but a serious candidate theory for complete unification—superstring theory—has already found.) This is physics in its traditional style. It is interesting to note that the paramount particle theorist of our time, the Nobelist Murray Gell-Mann, is in the forefront of the investigation of adaptive complex systems.

The astrophysicist Martin Rees, who is not known as much for any one specific accomplishment as for his polymathic understanding of the key cosmological questions, has remained at the forefront of cosmological debates. He is currently thinking about the possibilities of multiple universes, and how to take the weak form of the anthropic principle (as opposed to the strong form, which has marked religious overtones) and use it to illuminate that particular cosmological issue. He has had several important ideas on how stars and galaxies form, how to find black holes, and on the nature of the early universe. He is now trying to understand the mysterious "dark matter" which seems to fill intergalactic space—it is the gravitational pull of this dark matter which will determine whether our universe expands forever or eventually collapses to a "big crunch." He has always been interested in the broader philosophical aspects of cosmology. For instance: Why does our universe have the special features that allowed life to evolve? Are there other universes, perhaps governed by quite different physical laws?

Alan Guth, who began his scientific life as a particle physicist, has made what some consider to be the most important contribution to cosmology in a generation: the theory of inflation. In Guth's model, the very early universe underwent a period of rapid expansion; this accounts for, among other puzzles in big-bang theory, the present-day universe's puzzling homogeneity. Guth describes himself as taking a hard-nosed view of science, although his work is very often speculative. He is currently thinking about time travel: can wormholes in the fabric of space allow us to travel backward in time? Guth thinks the answer is no, but is fascinated by the fact that no one has been able to show that time travel is forbidden by the laws of physics.

The theoretical physicist Lee Smolin is interested in the problem of quantum gravity—of reconciling quantum theory with Einstein's gravitational theory, the theory of general relativity, to produce a correct picture of spacetime. He also thinks about creating what he calls a theory of the whole universe, which would explain its evolution, and he has invented a method by which natural selection might operate on the cosmic scale.

The theoretical physicist Paul Davies works in the fields of cosmology, gravitation, and quantum field theory, with particular emphasis on black holes and the origin of the universe. A prolific and influential popularizer of physics, he has written more than a dozen books. Here he presents the antireductionist agenda, and makes the case for moving both physics and biology onto "the synthetic path," recognizing the importance of the organizational and qualitative features of complex systems. He advocates a meeting of the minds between physicists and biologists, noting that complicated systems, whether biological or cosmological, are more than just the accretion of their parts but operate with their own internal laws and logic.

Chapter 15

MARTIN REES

"An Ensemble of Universes"

ALAN GUTH: *Martin Rees is my favorite theoretical astrophysicist. Whatever subject in astrophysics you ask him about, he's incredibly knowledgeable and incredibly helpful as well. If you ask him a question, he'll go on and on explaining in detail what is known about that subject. He's just marvelous.*

• • •

MARTIN REES *is an astrophysicist and cosmologist; Royal Society Research Professor at King's College, Cambridge; author of* Our Home Universe *(in press) and, with John Gribbin,* Cosmic Coincidences: Dark Matter, Mankind, and Anthropic Cosmology *(1989).*

MARTIN REES: The public is always interested in fundamental questions of origins. Just as they like dinosaurs, they're interested in cosmology. It's rather remarkable that the subjects which interest the public most consistently are sometimes so remote from everyday concerns. People who say that we have to make our work "relevant" to attract public interest are clearly on the wrong lines, because nothing could be less relevant than dinosaurs and cosmology.

Cosmology is exciting to the public because it's clearly fundamental, and this is a rather special time in the subject. For the first time, it's become a part of mainstream science, and we can address questions about the origin of the universe. We can talk about the details of what the universe was like when it was one second old. We can talk about even earlier stages, and ask basic questions; it's a very special and exciting era in the subject.

I would describe myself as an astrophysicist and cosmologist, in that order. An astrophysicist tries to understand individual objects, like galaxies, quasars, stars, and their evolution, whereas a cosmologist is concerned with the entire universe, not the contents of it. I try to span those two disciplines, which after all are very closely linked. I'm not a particularly good mathematician. My work tends not to be deductive system-building, as it were, but attempts to explain the phenomena.

When I started out, cosmology was primarily a theoretical subject, because there were essentially no data at all. It's only since the 1960s that we've known much beyond the fact that the universe is expanding. There have been exciting developments, to the extent that now we can talk in a quantitative way about the early stages of the universe, and there's been a tremendous extension in the range of cosmological issues we can discuss in a serious scientific way. Those issues used to be purely speculative, but now they're real science.

I haven't focused on any single fundamental question; I've tried to keep the big picture in mind, and I've been fortunate, because the subject is one in which a synthetic approach does often bear fruit. Data come in from optical telescopes, radiotelescopes, and spacecraft, and my colleagues and I try to put these together and make sense of them. It's like telling an engineer he's got to make something work meeting certain specifications. Nature gives us specifications, and we've got to use the laws of physics to see if we can "make something work," and make sense of these phenomena.

But there's always the nagging possibility that perhaps the laws of physics as we understand them now are inadequate. That's an extra motivation. The first reason for studying astronomy and cosmology is simply exploration, to discover what's out there. The second reason, which is what motivates astrophysicists, is to try to interpret what's out there and understand how the universe evolved, how the complexity of the present universe has emerged from the primordial simplicity. The third reason is that the cosmos is a laboratory that allows us to probe the laws of nature under conditions far more extreme than we could ever simulate in a terrestrial laboratory, and thereby to extend our knowledge of the fundamental laws of nature.

Another thing which interests me is the psychology of practitioners of the subject. Many people become strongly emotionally committed to their theories and defend them, almost like advocates, against contrary evidence. It's a real trauma for them to have to give their theories up. I've never been like that myself. I've always been quite happy to work almost simultaneously on two contradictory hypotheses, simply because if we don't really know what the explanation for something is, and we want to understand it, then exploring the consequence of different ideas is a good methodology. One's research may lead to a new test, or reveal a new contradiction. The scientific community collectively works like that, but not all individuals seem as content as I am to work simultaneously on two different theories.

One of the themes of my work is trying to understand extreme objects in the universe, objects that involve black holes, energetic outbursts, and so on. I'm associated with several ideas on quasars and the centers of galaxies. This subject is called high-energy astrophysics. In the last ten years, I've increasingly moved towards what you might call cosmogony. It's now feasible to learn not just about the present structure of the universe by surveying nearby galaxies, but about the early universe by looking at its distant parts, so that we're probing what the universe was like when galaxies were just forming, and even the pregalactic universe.

The most active areas in which I'm involved are how galaxies and galactic clusters formed, what the dark matter is, and whether the universe has enough material in it to cause it eventually to collapse or whether it will go on expanding forever. We don't yet have the answers, but I would expect that within the next decade we'll have a consensus view on some of those questions. I believe we'll understand more about how galaxies form, just as we now understand how stars form, and I hope we'll discover what the dark matter is. One of the embarrassing features of our current perception of the universe is that 90 percent of what it's made of is unaccounted for. This so-called missing material could be anything from very faint stars to exotic particles or black holes. Obviously, we can't understand the galaxies until we understand what makes up 90 percent of their mass.

We have good reason to believe that there's a lot of stuff in the universe which exerts a gravitational force but which we don't see. The simplest line of evidence comes from a disk galaxy, like our Milky Way, which is spinning. If you look at the outer parts of disk galaxies, you find that gas, way out, is orbiting surprisingly fast. It's orbiting faster than it would be if it was just feeling the gravitational pull of the stars you see. That's one line of evidence indicating that there must be a lot of dark matter holding these galaxies together. Other evidence comes from gravitational lensing and from the internal motions of clusters of galaxies. We believe that the dark matter is ten times as important gravitationally as what we see, and its nature is completely uncertain. But obviously the cosmogonic process—the origin of structure—is dominated by gravity, and therefore unless we know the nature of the stuff exerting most of the gravity, we're not going to have a definite answer to how the galaxies formed. The nature of the dark matter is one of the key uncertainties now.

If I was to say in one sentence what I'm trying to do, and what I suppose all cosmologists and cosmogonists are trying to do, it's simply to understand how the universe has evolved—over its fifteen-billion-year history—from a hot, compressed, amorphous fireball to its present state, in which we see galaxies and clusters of galaxies, and stars and planets, all displaying an enormous range of complexity of which we're a part. We want to understand the various stages in the emergence of structure: how the expanding universe developed condensations that turned into galaxies and galactic clusters, how stars formed in those, how the stars evolved, how the chemical elements were made, and how, on at least one planet, around at least one star, complex creatures evolved able to wonder about it all.

What's impressive is that we can address these questions at all. One reason we can is that in some respects the universe displays more simplicity than we had any right to expect. It displays simplicity in two senses. First, the large-scale structure of the universe is quite uniform and symmetric. There are all kinds of inhomogeneities on the scale of galaxies and clusters, but on the very large scale the universe is fairly uniform. Every bit has evolved and has the same history as every other bit, provided that by a "bit" we mean a "box" a few hundred million light-years across. In a broad-brush sense, the universe is smooth and homogeneous. When we look at a distant part of the universe, we're confident that we're seeing conditions as they were in our vicinity a long time ago. We could not assume that, if different parts of the universe had quite different histories.

The other remarkable feature is that the laws of physics are the same in all observed parts of the universe. When we take spectra of the light from distant quasars, the spectra indicate atoms just the same as those around us, and we believe that the laws established in the lab are adequate to explain everything in the observable universe, right back to when it was only a microsecond old. When we get earlier than a microsecond, the densities, energies, and pressures were so high that we have an uncertainty about the basic physical laws. After the first microsecond, the universe had expanded to where the densities were no higher than those we can achieve in the lab, and therefore we're likely to know the relevant physics.

There's also an added interest, because it's through the inferences we might be able to draw about the ultra-early universe—the first microsecond—that we can perhaps learn things about fundamental physics which we can't learn directly in the lab. Even in our biggest accelerators, we can't achieve the energies that particles possessed in the ultra-early universe. Also, many of the key properties of the universe—such as why it's expanding the way it is, why it has the simplicity and symmetry without which cosmology would be quite intractably difficult, and why it contains the observed ratio of matter to radiation—can't be understood without better knowledge of the first microsecond.

The microwave background radiation, discovered in 1965 by Arno Penzias and Robert Wilson, was the most important advance in cosmology since the late 1920s, when Edwin Hubble discovered that the universe was expanding. Hubble's discovery suggested that the universe had emerged from a compressed phase in the past, but there was then no evidence for that phase. Indeed, the steady-state theory, developed by some rather vocal Englishmen, held that such a compressed phase had never existed and that the universe had always been the same. It was the discovery of the background radiation that clinched the case for there having been a dense, hot, early stage of the universe, and almost all cosmologists became convinced fairly quickly. The resultant shift in cosmological opinion was almost as sharp as the concurrent shift in geophysical opinion in favor of continental drift—which was another formerly wildly speculative idea shown to be true. After the mid-sixties, almost everyone believed in the hot big-bang theory, of which this background radiation, now cooled to 2.7 degrees above absolute zero, was a fossil.

Since 1965, there's been a succession of more and more accurate measurements of the spectrum of this radiation, and of its angular distribution over the sky, because it's clearly a key cosmological probe. Two crucial discoveries were made. Nearly twenty years ago, the astrophysicist George Smoot measured our motion relative to the universe by finding that the background radiation, instead of being exactly the same temperature in every direction around us, was slightly hotter in one direction than in the opposite direction. This is because we and our entire galaxy are moving relative to the frame of reference defined by the large-scale universe, at a few hundred kilometers per second. Smoot made this discovery by flying his equipment on a U-2 spy plane and measuring the background radiation to a precision of more than one part in a thousand.

Smoot then went on to become one of the key people involved in the COBE satellite, which was launched in 1989 to investigate the radiation further. He was the PI, or principal investigator, for an instrument that looked for variations in the radiation temperature in different parts of the sky to a precision of one part in a hundred thousand. He found that the temperature was not completely uniform: some regions were slightly colder than others. The interpretation of this is that the early universe was not completely smooth. It's smooth in the sense that the surface of the ocean is smooth—a mean curvature, but with ripples superimposed on it. The "ripples" had been predicted to exist, as the seeds from which galaxies and clusters formed. Smoot's instrument onboard the COBE satellite was the first one sensitive enough to have found these fluctuations.

Had they not been found at that level of sensitivity, persons like myself would have been deeply disconcerted, because we all believed that the galaxies, clusters, and superclusters had formed by gravitational instability—a process whereby any part of the early universe that was slightly denser than average would lag behind as the universe expanded, and would eventually condense out. Galactic clusters and superclusters could not have condensed out by the present time unless inhomogeneities already existed in the early universe, with an amplitude that would imprint one-part-in-a-hundred-thousand fluctuation in the microwave background. That was the level theorists knew one had to shoot for in doing this experiment, and that was the level achieved by Smoot's instrument on the COBE satellite.

I'm interested in what general properties our universe had to have in order to develop complexity. One of the obvious requirements is a force like gravity, which allows structures to condense, via instabilities, in an initially featureless universe. But, ironically, the weaker gravity is, the better the chances of a complex universe developing, because if gravity were so strong that it crushed things the size of complex organisms, there would be bleak prospects for evolution. Indeed, if gravity were much stronger, the lifetimes of stars would also be very short, and this would allow less time for complexity to emerge via any evolutionary process. If the force of gravity didn't exist, no cosmic structures would ever have condensed, but the weaker it is the grander its manifestations are. It's because gravity is so weak that stars and galaxies are so huge, on the scale of ordinary phenomena. It's interesting to try to quantify this, and see if we can understand why gravity should be so weak.

The general idea of the emergence of complexity is very relevant here, because gravity has the unusual property of allowing an initially featureless universe to develop structure. Gravity leads to instabilities, and pulls material together to form galaxies and stars. As stars lose energy, they get even hotter in their centers and more compact; eventually, nuclear-fusion reactions ignite inside them, allowing the temperature contrasts between stars, planets, and the dark night sky that are essential—as Prigogine and others have taught us—for the "nonequilibrium thermodynamic" processes that built up complex molecules and life. So gravity drives things ever further from equilibrium and allows the disequilibrium, which is the prerequisite of any kind of complexity, to develop from an amorphous early universe. This is the kind of process we're trying to understand quantitatively. Another development in the last few years is the possibility of doing realistic simulations of gravitational clustering, gas dynamics, and so forth, to explore how a structureless universe can evolve.

I've also ventured into more speculative topics, like whether physicists might by accident destroy the universe by doing a particular sort of experiment. This issue arose because of ideas stemming from Alan Guth's inflationary-universe theory. The whole idea of the inflationary universe requires that even empty space (what physicists call "the vacuum") had unusual properties in very early times and underwent what's called a phase transition—something like what happens when water freezes. Some people—the physicist Sidney Coleman was one of the first to make this point—suggested that our present vacuum may not be in the lowest possible energy state. Space might therefore undergo a further phase transition to a different kind of vacuum state, in which the laws of physics would be changed. All particles as we know them, and everything we see around us, would be destroyed. Our present vacuum may be, so to speak, supercooled, as very pure water can be supercooled without undergoing the phase transition to ice; and, just as the insertion of a speck of dust makes supercooled water freeze suddenly, maybe some trigger could transform the whole of space into some other quite different state. Could physicists, by an experiment done in an accelerator, trigger this effect by inadvertently producing a bubble of the new vacuum, which would then expand at the speed of light and engulf the universe?

This might appear absurd, but it's easy to think of ways in which we've produced conditions that have never existed naturally any-where. For instance, there was never anything in the universe colder than 2.7 degrees above absolute zero—the present temperature of the microwave background—until we made refrigerators (unless, that is, there's intelligent life elsewhere). The kind of thing that might create "dangerous" conditions would be a collision between very-high-energy particles in a big accelerator; such a collision might cre-ate a big local energy density of just the kind that might trigger a phase transition.

With the Dutch astrophysicist Piet Hut, I wrote a paper address-ing the question of whether accelerators could create concentrations of energies that had never existed anywhere in the universe since the big bang itself. Our conclusion was quite reassuring. We calculated the collision rate between cosmic-ray particles—which are particles that move, at very low densities, in interstellar space, at very close to the speed of light. We worked out the most energetic collisions that ever happened in our part of universe, and we discovered that these would have been substantially more energetic than any conceivable event that could occur in an accelerator. That's reassuring. It means you'd have to go a long way beyond the collision energies expected in supercolliders before there was any risk of Doomsday.

I'm also trying to bring into a scientific context the concept of an ensemble of universes, each with different properties. These ideas are associated with many people, but I'll mention only the Russian physicist Andrei Linde, who proposes chaotic and eternal inflation—that is, the idea that new universes can sprout from old ones, or can inflate into a new domain of spacetime inside black holes. He and others have argued that our universe is just one element in an infi-nite ensemble. Different universes in this ensemble may be governed by entirely different physical laws, numbers, and dimensions. Some may have very strong gravitational force, some may have no gravity, some may have different kinds of particles. If that's a possibility, then this concept of an ensemble, which I prefer to call a meta-universe, gives a scientific basis to anthropic reasoning—the idea that it's not a coincidence that we find ourselves in a universe where conditions are somehow attuned for the development of complexity. If all possi-ble universes governed by all possible laws exist, then obviously it oc-casions no surprise that some of them will have laws of nature that allow complexity, and then it's no coincidence—and, indeed, inevitable—that a universe like ours exists, and, of course, that's the one we're in. This suggests the idea of "observational selection," as it were, of universes. I take this seriously. There's an ensemble of universes. Insofar as one can put a "measure"—in the mathematical sense—on relative numbers of universes, most will be stillborn, in the sense that there would be no complexity evolving within them. Some, contrariwise, may have vastly greater potentialities than our own, but these are obviously beyond our imaginings.

I have substantial confidence in talking about the universe back to when it was a microsecond old; I have as much confidence in the relevant theories as I have in inferences about the early history of the earth from geophysics or paleontology. The level of evidence and the nature of the argument are similar—indeed, the cosmological evidence is rather more quantitative. But when we get back into the first microsecond, we confront important ideas, like inflation and phase transitions, which in some form will be part of the eventual correct world picture. The trouble is that we don't yet know enough about the extreme physics to be able to predict anything very quantitatively. But the new concepts certainly expand our perspective, by admitting the possibility of an entire ensemble of other universes with different properties. We have to then distinguish different definitions of "universe." You can mean by "universe" what we observe—a region some fifteen billion light-years across; you could define it as larger than that—as the domain from which light will eventually be able to reach us; or you could define it as the grand ensemble, which contains all possible universes, governed by all possible physical laws. It's the last concept—the meta-universe—which I find the most fascinating, and which I believe is just coming within the scope of serious scientific discourse.

Rather than use the phrase "anthropic principle," I would prefer to talk about "anthropic reasoning." This is the general line of argument that some features of the universe are a prerequisite for the existence of observers, and so we shouldn't seek a basic explanation of those features: they're just a function of the fact that we're here. In one sense, anthropic reasoning is obvious and quite banal; we don't bother wondering why we're in a special place in the universe, near a star like the sun, and not in a random place in intergalactic space. Nor do we wonder about why we're living in the universe when it's fifteen billion years old rather than in the first few seconds, because for us to exist the universe had to cool down, and a long chain of prior evolution plainly must have occurred.

Some people have tried to take anthropic reasoning further, by claiming that it's somehow mandatory that any basic laws of nature must permit conscious observers. I find this view hard to take seriously. The status of anthropic reasoning depends very much on the nature of the basic laws. If these laws—that is, the relative strengths of gravity and the other fundamental forces, the masses, spins, and charges of the elementary particles, and so forth—are, in a sense, accidents of the way our universe cooled down, then you can perfectly well imagine universes where the laws are different and which are not propitious for life. All these universes might exist, and we happen to be in the one that has the "right" conditions. There's nothing remarkable about that.

On the other hand (and here I would sympathize more with the line taken by the late physicist Heinz Pagels), if the fundamental laws of nature are unique—if it turns out that the laws of physics could not have any other form anywhere, and some unique equation tells us the strength of the forces and the masses of the particles—then it would seem just a brute fact, or luck or providence, according to your perspective, that those unique and simple laws allow complexity to evolve. I'd be astonished at this outcome, but my reaction would be rather like—to make an analogy—my amazement at the fact that you can write down a simple algorithm for something as complex as the Mandelbrot set, with its infinite depths of structure. It is indeed amazing, but that's just mathematics; and, similarly, there may indeed be unique fundamental physical laws that just happen to have such incredibly rich consequences.

If the laws of nature are unique, then there's no room for anthropic selection, because the laws are just "given." Either you accept the laws, and their remarkable consequences, just as a brute fact, or you go all the way with the "strong anthropic principle." But if there's an ensemble of universes that cool down differently, then some have conditions propitious for life, and others are short-lived; they will be too cold, too empty, and so forth. Then there *is* room for straightforward anthropic selection. And we are perforce in one that's hospitable enough to allow the requisite complexity.

A reason for downplaying anthropic arguments is that physicists would do well not to believe them too strongly. Obviously, many features of the universe we can't yet account for will be explained by straightforward physical arguments. If people believed that some features of the universe were not fundamental but just accidents, resulting from the particular way our domain in the meta-universe cooled down, then they'd be less motivated to try to explain them. When I interviewed Steven Weinberg for a radio documentary about ten years ago, he made this point—that it would be best that physicists not believe in the anthropic principle, because otherwise they wouldn't be so motivated in seeking a unified theory, and if they didn't seek it they certainly wouldn't find it. These new concepts of a meta-universe (or ensemble of universes) bring anthropic selection closer to the mainstream of scientific discourse.

LEE SMOLIN: I met Martin Rees only recently, during a visit to Cambridge University. Of course, I'd heard about him for many years, as he's admired by a great many people. He's certainly one of the most influential people working in astrophysical and cosmological theory, and after some discussions with him it was obvious to me why: he is simultaneously open to new ideas and suggestions and careful and rigorous in his response and criticisms. Also, it's difficult to suggest an idea about the evolution of structure in the universe or the formation of the galaxies that he hasn't thought of or played with or perhaps even written about at some time. He's also great fun to talk with, and as far as I could tell completely without pretention. It's not fun to hear some people criticize one's ideas, because they turn such discussions into something competitive, but I can say that I really enjoyed hearing his criticisms of some ideas of mine. He didn't believe them, but he'd thought about them carefully, and he told me exactly where he thought they were most likely to go wrong.

Much of the credit for what I like to think of as the discovery that the laws of nature are special in ways that allow the universe to be very structured is due to him. This idea was suspected initially by an earlier generation—particularly by P.A.M. Dirac, Fred Hoyle, and Robert Dicke. But it's my understanding that it was really Martin, together with a younger colleague, Bernard Carr, who assembled all the evidence for the specialness of the laws of nature. The result was a paper they published in *Nature* which has had an enormous influence on all those who think about the anthropic principle. More than one book has been written by expanding their article. But what I think is most important is that they've made the case for the special-ness of the laws of nature strong enough so that those of us, like my-self, who aren't attracted to the anthropic principle have to take it seriously. Then the question is, If we don't accept our own existence as the explanation for why the universe is so special, can we find an-other explanation?

If I can use him to say something more generally, there's some-thing truly wonderful about the English tradition in astronomy and physics that we in America could learn a lot from. There's no country in the world that has had such a collection of inspired originators of cosmological and astronomical ideas. In this century there has been Arthur Eddington, Fred Hoyle, Dennis Sciama, Roger Penrose, Stephen Hawking, and Martin Rees himself, and there are others not as well known. There's a way in which these people have been edu-cated to work with the highest standards of rigor and honesty, and then allowed to develop their ideas in an atmosphere much freer and more tolerant of individuality, and even eccentricity, than the Ameri-can scene. The American scene is larger, and in terms of money we're better supported, but there's something unhealthy about the way in which we're so often worrying about how the National Science Foundation and the community will respond to our grant proposals. Perhaps I'm naive, but I have the impression that the British seem, at least up until now, to have avoided this overbureaucratization of sci-ence.

It's also only England that could have produced scientists like Jim Lovelock or the physicist and philosopher Julian Barbour, who stay at home, unconnected to any university, but do original and impor-tant work that wins the respect of their less courageous colleagues in the universities. Perhaps the point is that the English have never for-gotten that in the end the advances of science are made by creative personalities, so that the best way to advance science is to give peo-ple the best possible education intellectually and morally—I say "morally" because I think science works because scientists practice an ethics of honesty and tolerance—and then give as much freedom as possible to those who show themselves to be creative. This is something I think we need to think about more in the United States.

NICHOLAS HUMPHREY: Martin and I are friends, and I turn to him when I want to know about physics or cosmology. Sometimes I think he's a bit too levelheaded. Martin and I disagree, for example, about the strong anthropic principle. I think the strong anthropic principle is wonderful—not necessarily true, but wonderful—but Martin has no time for it.

ALAN GUTH: Martin Rees is my favorite theoretical astrophysicist. Whatever subject in astrophysics you ask him about, he's incredibly knowledgeable and incredibly helpful as well. If you ask him a question, he'll go on and on explaining in detail what is known about that subject. He's just marvelous. It's a little difficult to put one's finger on Martin's accomplishments, because they're so widespread. He's probably written several hundred papers in the astrophysics literature, making important contributions to almost every aspect of the subject.

Chapter 16

ALAN GUTH

"A Universe in Your Backyard"

LEE SMOLIN: *The idea of inflation has probably been the most influential idea in cosmology in the last fifteen years, and it's Alan's idea. It's an idea that hasn't entirely convinced me, and I'm not alone in this, but it's had an enormous effect on everybody's thinking.*

• • •

ALAN GUTH is a physicist; Victor F. Weisskopf Professor of Physics at MIT; author of The Inflationary Universe *(in press).*

ALAN GUTH: Cosmology has very much become an observational science; it's no longer people sitting back in armchairs inventing unfounded theories about what the universe might look like. Observations are being made all the time: observations of the distribution of galaxies in the universe, observations of the microwave background radiation and the nonuniformities in that radiation; estimates of the mass density of the universe; estimates of the age of the universe, based on a variety of different techniques.

All that has an impact on the kinds of theories of the universe which are viable. In 1980, I developed the idea of the inflationary universe. It was a new theory of how the big bang might have begun. It's a theory consistent with the standard big-bang picture, which is one of the reasons it's become as well accepted as it has. It doesn't require people to throw out what was believed previously about cosmology. But it adds a lot. It adds a whole story about what happened during the first fraction of a second of the universe, a time period that had not been explored before. It answers a number of questions left open by the standard big-bang model. The inflationary universe is a theory about reality. I, and probably most physicists, regard reality as a genuine physical reality, a reality influenced by people only insofar as we can reach and move things and so on. Reality exists independent of people. The goal of the physicist is to understand that reality.

One of the most amazing features of the inflationary-universe model is that it allows the universe to evolve from something that's initially incredibly small. Something on the order of twenty pounds of matter is all it seems to take to start off a universe. This is very different from the standard cosmological model. Before inflation, the standard model required you to assume that all the matter that exists now was already there at the beginning, and the model just described how the universe expanded and how the matter cooled and evolved. Given the inflationary model, it becomes very tempting to ask whether, in principle, it's possible to create a universe in the laboratory—or a universe in your backyard—by man-made processes.

The first question to look at is what would happen if you had a small patch of inflationary universe in the midst of our universe, never mind how it might have gotten there. Let's pretend that it exists, and ask how it evolves. It turns out that if this patch is big enough, it will grow to become a new universe, but it does this in a very strange way. It doesn't—and this is very important for environmental purposes—displace our universe. Instead, the patch forms a wormhole and slips through it. From our universe, it always appears very small and looks more or less like an ordinary black hole. But on the inside, the new universe is expanding and can become arbitrarily large, creating new space as it grows. It can easily become large enough to encompass a universe like the one we see. In a very short length of time, a small fraction of a second, it completely pinches off from our universe and becomes a totally isolated new universe.

Inflationary cosmology is a new twist on the big-bang theory. It doesn't in any way do away with the big-bang theory. It's completely consistent with everything that's been talked about in terms of the big-bang model. What it does is change our conception of the history of the first small fraction of a second of the big bang. According to the new theory, the universe during this sliver of time underwent a period of inflation, a brief era of colossal expansion.

There are two key features that are different in inflationary cosmology from the standard big bang. One is that the inflationary model contains a mechanism by which essentially all the matter in the universe can be created during the brief period of inflation. In the standard big-bang model, by contrast, it was always necessary to assume that all the matter was there from the beginning, and there was no way to describe how it might be created. By the way, the inflationary production of matter is consistent with the principle of energy conservation, even though it can literally produce a universe from almost nothing. Energy is still conserved—this is all calculated in the context of standard classical general relativity. The unusual feature is that gravity plays a major role in the energy balance. It turns out that the energy of a gravitational field—any gravitational field—is negative. During inflation, as the universe gets bigger and bigger and more and more matter is created, the total energy of matter goes upward by an enormous amount. Meanwhile, however, the energy in gravity becomes more and more negative. The negative gravitational energy cancels the energy in matter, so the total energy of the system remains whatever it was when inflation started—presumably something very small. The universe could, in fact, even have zero total energy, with the negative energy of gravity precisely canceling the positive energy of matter. This capability for producing matter in the universe is one crucial difference between the inflationary model and the previous model.

The other big difference is the ability of the inflationary theory to explain several prominent features of our universe which remain unexplained in the standard big-bang model. Take, for example, the large-scale uniformity of the universe. When we look out to great distances, it appears that the universe is remarkably uniform. The best evidence for this comes from the oldest thing we can see—the cosmic microwave background radiation, a kind of afterglow of the big bang itself. When we look at this background radiation, we're seeing a snapshot of what the universe looked like when that radiation was released—something that happened only a few hundred thousand years after the big bang—and it's telling us that the universe was then incredibly uniform.

In the context of the standard big-bang model, that was always a mystery. The early universe was so large that there wasn't nearly enough time for light to travel across it in the time available. We can imagine, for example, observing the microwave radiation from two opposite directions in the sky, and then we can use the big-bang theory to trace each of the two microwave beams back to its source. When the radiation was released, the two sources were separated from each other by a distance about a hundred times larger than the total distance that light could have traveled up until that time. Since we believe that nothing can travel faster than light, it means that the point on one side of the universe had no way of being influenced by what was going on at the opposite point, but somehow they managed to be at the same temperature at the same time to the extraordinary precision of a few parts in a hundred thousand. The standard big-bang theory could account for this uniformity only by assuming, without explanation, that the universe started out incredibly uniform.

The inflationary model, on the other hand, posits a short period in the very early universe during which the universe expanded far, far faster than in the standard cosmology. This implies that the early universe was far smaller than people had previously thought. There was plenty of time for this microscopic proto-universe to come to a uniform temperature before inflation began, and then inflation magnified this very small region to become large enough to encompass the observed universe. The large-scale uniformity of the universe is therefore no longer a mystery, but can now be understood as the natural consequence of cosmic evolution. To account for the observed degree of large-scale uniformity, we must assume that the universe expanded during the inflationary era by at least a factor of a trillion trillion. It's quite likely that the expansion factor was much larger than this stupendous number, but we have no way of knowing how much the universe actually inflated.

I've recently been working on wormholes and on the question of whether it's in principle possible to create "a universe in your backyard." A few years ago I worked with Steven Blau and Eduardo Guendelman to figure out what would happen if there were a region of an inflating universe in the midst of our universe. We found that the question could be answered very cleanly and unambiguously, since the behavior is determined by general relativity. The only new ingredient for this problem is an idea from particle physics about a certain kind of matter called a "false vacuum," which is the driving force behind inflation. We discovered that a large enough region of false vacuum would create a new universe, which, as I described earlier, would rapidly disconnect from ours and become totally isolated.

The next question, which turns out to be much harder, is what does it take to produce this small region of false vacuum—to start everything going? Since the mass density of the false vacuum is approximately 10^{60} times larger than the density of an atomic nucleus, it would certainly not be easy. There's no technology in the present or the foreseeable future that would allow us to do this sort of thing. Nonetheless, one can talk about the physics of universe creation as a matter of principle, and I find it a very interesting question.

I'm going to imagine that somebody can make a false vacuum and learn to manipulate these extraordinary energy densities. But then there's still another problem. As you start to collect this material, its own gravitational force is so strong that it tends to collapse into a black hole. The formation of a black hole can be prevented only by starting the material expanding at a very high speed. We found that if the region is to expand fast enough to produce a new universe, it must begin from what in technical terms is called an initial singularity—also known as a white hole. A white hole is essentially the opposite of a black hole: while matter can fall into a black hole but cannot escape, matter is ejected from a white hole but cannot enter it.

The instant of cosmic creation in the big-bang theory is an example of a white hole, but certainly nobody has ever seen a white hole, and nobody knows how to make one in the laboratory. So if you ask whether a new universe can *in principle* be created in the laboratory, the answer, according to classical general relativity, is no, since such a creation requires a white hole. But classical general relativity is not the final word. The evidence is overwhelming that we live in a quantum universe—a universe that isn't governed by deterministic classical laws. We've found that quantum theory is absolutely essential for understanding molecules, atoms, and subatomic particles, and physicists firmly believe that quantum theory is also essential to an understanding of the true nature of gravity. Unfortunately, however, there are very complicated technical problems in trying to construct a quantum theory of gravity. The riddle of quantum gravity is perhaps solved by the superstring theory, but that theory is so poorly understood that it hasn't yet been used to answer any of the central questions that quantum gravity is expected to address.

While classical physics implies that a universe can't be created without a white hole, there's a possibility that quantum effects could make it easier. Edward Farhi, Jemal Guven, and I attempted to study the quantum question using an approximate formulation of quantum gravity that's much more tractable than superstring theory. We discovered two things. First, we found that one of the standard approximations to quantum gravity led to inconsistencies and had to be modified to obtain any answer at all. Second, we found that if we believed our modified rules of quantum gravity, then it *is* in principle possible to create a universe in the laboratory without starting from a white hole. The procedure isn't guaranteed to succeed, but in the context of quantum mechanics we were able to estimate a probability for success. Since our calculations relied on a modification of an approximation that was uncertain in the first place, we found it reassuring that Willy Fischler, Daniel Morgan, and Joseph Polchinski obtained the same results with a different method. The probability of success was found to depend crucially on the energy density of the false vacuum. If it's at a scale typical of what particle physicists call "grand unified theories," then the probability would be outlandishly small. On the other hand, it's conceivable that the energy level associated with the false vacuum might be a thousand times larger than those of the grand unified theories, and then the probability of successful universe production would be high.

Our calculations remain somewhat tentative, however, as the uncertainties of quantum gravity haven't been overcome. Since synthetic universe creation is well beyond the range of experiment, the only chance for discovering within our lifetimes whether it's possible would be the development of detailed theories of quantum gravity and the behavior of matter at extremely high energies. Those two challenges are linked, since the gravitational interactions of elementary particles become significant only at extraordinarily high energies.

An interesting aspect of the universe-creation work was the role of wormholes—elongated tubes of space that can in principle connect one universe to another, or a part of a universe to a distant part of the same universe. In the universe-creation scenario, the child universe is initially connected by a wormhole to its parent, although the wormhole pinches off in about 10^{-35} seconds. The same kinds of wormholes are also relevant to the question of whether the laws of physics allow the possibility of time travel.

The question of time travel hinges on the lifetime of the wormholes. For time travel to work, one needs to have a stable wormhole—a wormhole that can be built large and exist for a long time, so that you could travel through it. The scenario would begin with the construction of a wormhole linking our universe to itself, whenever it becomes technologically feasible. Then the aspiring time-traveler would keep one entrance of the wormhole alongside her as she evolves normally into the future. She must keep the entrance moving at near the speed of light, but it can travel in a circle, so that it returns periodically. Years or millennia later, she or her descendants would be able to return to the time at which the wormhole was constructed by traveling through the wormhole.

The laws of physics, however, are not very cooperative with wormhole-transportation engineers. The rapid collapse of the wormhole in child-universe production is characteristic. In fact, if the wormhole is constructed from any "normal" material, it will collapse before *anything* can go through it. To hold the wormhole open requires a material with a negative energy density. There's room for hope, however, since relativistic quantum theories are known to allow the existence of regions of negative energy density. The size and duration of such regions are limited, however, so no one has yet designed a theoretically traversable wormhole. On the other hand, no one has proven it impossible.

People might wonder whether it makes any sense at all to be playing with theories that involve numbers such as 10^{-35} seconds. "How can you assign value or meaning to a number like that?" some may ask, since it's so far beyond direct experience. One of the amazing things about science, though, is the spectacular success we've had in extrapolating mathematical relationships. When the equations of electricity and magnetism were assembled by Maxwell in 1864, for example, they were based on tabletop experiments, with distances ranging from centimeters to meters. Today we successfully use these same equations to describe phenomena ranging from the size of atomic nuclei to the size of the visible universe. Obviously, however, one cannot claim that such extrapolations are always valid. When Newton's laws of motion are extrapolated to half the speed of light, they are found to be wrong! While large extrapolations aren't necessarily trustworthy, I would claim that they are always worth exploring. Special relativity was discovered, in fact, when Einstein attempted to extrapolate Newton's laws to near the speed of light. What would it look like to ride on a light wave? Einstein asked himself. Today physicists are similarly asking what it would look like to view the universe 10^{-35} seconds after its birth. It's speculative, but it's also intriguing, and we hope that it's productive.

I tend to take a rather hard-nosed point of view as to the underlying nature of the universe. The universe exists as a physical object, and physicists and other scientists are making a lot of progress in trying to understand the rules by which it works. It's important in science, and in life, to recognize that at any given time there will always be some questions you can't answer. You continue to try to answer them, but you shouldn't be surprised if you find you're incapable of answering them.

LEE SMOLIN: The idea of inflation has probably been the most influential idea in cosmology in the last fifteen years, and it's Alan's idea. It's an idea that hasn't entirely convinced me, and I'm not alone in this, but it's had an enormous effect on everybody's thinking. The idea came from attempts to understand some very difficult problems about the universe as a whole; particularly why it's so symmetric, why it's not much, much more disordered than it is. If we believe in the standard theories of cosmology, and believe that the big bang was the first moment of time, then there was not enough time from that initial moment to the moment we see when we look at the cosmic microwave background radiation for the different parts of the universe to have interacted with each other and come to the same state. Every part of the universe we can see has the same temperature to a few parts in a hundred thousand. The idea of inflation was invented to explain that and other puzzles.

There are two ways to talk about inflation. You can say that at some very, very early moment the universe expands exponentially fast, growing by many, many powers of ten; *or* you can say that time slows down extraordinarily during this inflationary period. Both have the same effect; the period allows all the different parts of the universe we see to have been in communication.

Another very interesting thing about the idea is that it makes a prediction, which is that omega is precisely equal to one. Omega is a measure of the density of matter in the universe. There's a certain density of matter which will eventually cause the universe to collapse back—this collapse would be brought about by the matter's mutual gravitational pull. We can compute, since we know the speed at which the expansion is happening, how much matter there needs to be to stop the expansion. The ratio of the density of the matter that's actually here in the universe to this critical density is called omega. It is *the* critically important quantity in cosmology. The theory of inflation predicts that omega should be equal to one, which means that the universe is precisely balanced at the border between those that collapse and those that expand forever. A fraction more matter and it must collapse, a fraction less and it will at some point enter an eternity of rapid, runaway expansion. Therefore the inflationary theory is subject to experimental tests, observational tests. These tests are going to occur in the next ten or fifteen years.

MARTIN REES: Inflation has been the stimulus for a great deal of the cosmological discussion about the ultra-early universe. There have been various fashions—old inflation, new inflation, chaotic inflation—and the actual details are still uncertain, in that we can't, for instance, calculate exactly when it happened in the universe, or how the fluctuations arose which evolved into galaxies, clusters, and superclusters, because the answers depend on uncertain physics. It's an exciting new possibility, however, that the small temperature fluctuations first detected by COBE, and now by about ten other experiments, may reveal, spread across the sky, the imprint of physical processes that occurred when the entire observed universe was squeezed smaller than a golfball. The exotic physics of this ultra-early era, when quantum-uncertainty effects were important on the cosmic scale, is now accessible to observations, and speculations can soon be at least constrained. But most cosmologists would bet fairly high odds that the idea of inflation, which was first very clearly stated in Alan Guth's paper, is going to be an element of any correct theory about the early universe. It was one of the developments that made it possible to talk seriously about not just the first second but the first 10^{-36} seconds of the universe.

It was fortunate that Alan Guth did his work at the same time that another idea came into fashion, which was the theory that we could understand why the universe contains matter and not antimatter in terms of some asymmetry, some favoritism for matter over antimatter in the early universe; it's no good having a scheme that can inflate the universe to enormous dimension if it's not possible to create matter to fill that large universe. Guth brought these two ideas together in his inflationary cosmology.

Chapter 17

LEE SMOLIN

"A Theory of the Whole Universe"

MURRAY GELL-MANN: *Smolin? Oh, is he that young guy with those crazy ideas? He may not be wrong!*

· · ·

LEE SMOLIN *is a theoretical physicist; professor of physics and member of the Center for Gravitational Physics and Geometry at Pennsylvania State University; author of* The Life of The Cosmos: A New View of Cosmology, Particle Physics, and the Meaning of Quantum Physics *(1995).*

LEE SMOLIN: What is space and what is time? This is what the problem of quantum gravity is about. In general relativity, Einstein gave us not only a theory of gravity but a theory of what space and time are—a theory that overthrew the previous Newtonian conception of space and time. The problem of quantum gravity is how to combine the understanding of space and time we have from relativity theory with the quantum theory, which also tells us something essential and deep about nature. If we can do this, we'll discover a single unified theory of physics that will apply to all phenomena, from the very smallest scales to the universe itself. This theory will, we're quite sure, require us to conceive of space and time in new ways that take us beyond even what relativity theory has taught us.

But, beyond even this, a quantum theory of gravity must be a theory of cosmology. As such, it must also tell us how to describe the whole universe from the point of view of observers who live in it—for by definition there are no observers outside the universe. This leads directly to the main issues we're now struggling with, because it seems very difficult to understand how quantum theory could be extended from a description of atoms and molecules to a theory of the whole universe. As Bohr and Heisenberg taught us, quantum theory seems to make sense only when it's understood to be the description of something small and isolated from its observer—the observer is outside of it. For this reason, the merging of quantum theory and relativity into a single theory must also affect our understanding of the quantum theory. More generally, to solve the problem of quantum gravity we'll have to invent a good answer to the question: How can we, as observers who live inside the universe, construct a complete and objective description of it?

Most of my work as a scientist has been directed to the problem of quantum gravity. I like working on this problem a great deal, especially as it's the only area of physics I know of where one is daily confronted by deep philosophical problems while engaged in the usual craft of a theoretical physicist, which is to make calculations to try to extract predictions about nature from our theoretical pictures. Also, I like the fact that one needs to know a lot of different things to think about this problem. For example, it's likely that quantum gravity may be relevant for understanding the observational data from astronomy, and it's also likely that the new theory we're trying to construct will make use of new mathematical ideas and structures that are only now being discovered. So although I've worked almost solely on this problem for almost twenty years, I've never been bored.

I have days in which I spend the morning working on a calculation, to check an idea I had the night before, and then I'll go to a lunch seminar, where I hear astronomers discuss the latest evidence for some crucial question, like how much dark matter there is. Then I spend the afternoon studying the paper of a friend who's a pure mathematician, after which I meet a philosopher for dinner and continue an argument we're having on the nature of time. And what's wonderful is the way that these different subjects, which until recently were disconnected from one another, often seem to illuminate one another. Of course, sometimes it's not so ideal; teaching and bureaucracy take up a lot of time—although in reasonable doses, I must say. I love teaching also. But there are really many days when I feel very fortunate and can't imagine that I'm being paid to live like this.

For the last eight years or so—really, it doesn't seem so long!— I've been working with several friends on a new approach to combining relativity and quantum theory. We call this approach "nonperturbative quantum gravity." It's enabling us to investigate the implications of combining general relativity and quantum theory more deeply and thoroughly than was possible before. We aren't yet finished, but we're making progress steadily, and recently we've got the theory well enough in hand that we've been able to extract some experimental predictions from it. Unfortunately, the predictions we've been able to make so far can't be tested, because they're about the geometry of space at scales twenty orders of magnitude smaller than an atomic nucleus. But this is further toward a solution to the problem than anyone has gotten before—and, I must say, further than I sometimes expected we'd be able to go in my lifetime.

In this work, we've been combining a very beautiful formulation of Einstein's general theory of relativity discovered by my friend Abhay Ashtekar with some ideas about how to construct a quantum theory of the geometry of space and time in which everything is described in terms of loops. That is, rather than describing the world by saying where each particle is, we describe it in terms of how loops are knotted and linked with one another. This approach to quantum theory was invented by another friend—Carlo Rovelli—and myself, and also by the very interesting Uruguayan physicist Rodolfo Gambini.

The main result of this work is that at the Planck scale, which is twenty powers of ten smaller than an atomic nucleus, space looks like a network or weave of discrete loops. In fact, these loops are something like the atoms out of which space is built. We're able to predict that—just as the possible energies an atom can have come in discrete units—when one probes the structure of space at this Planck scale, one finds that the possible values the area of a surface or the volume of some region can have also come in discrete units. What seems to be the smooth geometry of space at our scale is just the result of an enormous number of these elementary loops joined and woven together, as an apparently smooth piece of cloth is really made out of many individual threads.

Furthermore, what's wonderful about the loop picture is that it's entirely a picture in terms of relations. There's no preexisting geometry for space, no fixed reference points; everything is dynamic and relational. This is the way Einstein taught us we have to understand the geometry of space and time—as something relational and dynamic, not fixed or given a priori. Using this loop picture, we've been able to translate this idea into the quantum theory.

Indeed, for me the most important idea behind the developments of twentieth-century physics and cosmology is that things don't have intrinsic properties at the fundamental level; all properties are about relations between things. This idea is the basic idea behind Einstein's general theory of relativity, but it has a longer history; it goes back at least to the seventeenth-century philosopher Leibniz, who opposed Newton's ideas of space and time because Newton took space and time to exist absolutely, while Leibniz wanted to understand them as arising only as aspects of the relations among things. For me, this fight between those who want the world to be made out of absolute entities and those who want it to be made only out of relations is a key theme in the story of the development of modern physics. Moreover, I'm partial. I think Leibniz and the relationalists were right, and that what's happening now in science can be understood as their triumph.

Indeed, in the last few years, I've also realized that the relational point of view can inspire ideas about other problems in physics and astronomy. These include the basic problem in elementary-particle physics, which is accounting for all the masses and charges of the fundamental particles. I've come to believe that this problem is connected as well to two other basic questions that people have been wondering about for many years. The first of these is: Why are the laws of physics and the conditions of the universe special in ways that make the universe hospitable for the existence of living things? Closely related to this is the second question: Why, so long after it was formed, is the universe so full of structures? Beyond even the question of life, it's a remarkable fact that our universe seems, rather than having come to a uniform and boring state of thermal equilibrium, to have evolved to a state in which it's full of structure and complexity on virtually every scale, from the subnuclear to the cosmological.

The picture that emerges from both relativity and quantum theory is of a world conceived as a network of relations. Newton's hierarchical picture, in which atoms with fixed and absolute properties move against a fixed background of absolute space and time, is quite dead. This doesn't mean that atomism or reductionism are wrong, but it means that they must be understood in a more subtle and beautiful way than before. Quantum gravity, as far as we can tell, goes even further in this direction, as our description of the geometry of spacetime as woven together from loops and knots is a beautiful mathematical expression of the idea that the properties of any one part of the world are determined by its relationships and entanglement with the rest of the world.

As we began to develop this picture, I also began to wonder whether the basic philosophy behind it might extend to other aspects of nature, beyond just the description of space and time. More precisely, I began to wonder whether the world as a whole might be understood in a way that was more interrelated and relational than in the usual picture, in which everything is determined by fixed laws of nature. We usually imagine that the laws of nature are fixed, once and for all, by some absolute mathematical principle, and that they govern what goes on by acting at the level of the smallest and most fundamental particles. There are good reasons why we believe that the fundamental forces should act only on the elementary particles. But in particle physics we have been making another assumption as well: that there are mechanisms or principles that pick out which laws are actually expressed in nature, and that these mechanisms or principles also work only at enormously tiny scales, much smaller than the atomic nucleus; an example of such a mechanism is something called "spontaneous symmetry breaking." Given that the choice of laws makes a great difference for the universe as a whole, it began to seem strange to me that the mechanisms that choose the laws should not somehow be influenced by the overall history or structure of the universe at very large scales. But, for me, the real blow to the idea that the choice of which laws govern nature is determined only by mechanisms acting at the smallest scales came from the dramatic failure of string theory.

Like many of the young people trained in elementary-particle physics in the 1970s and '80s, I had great hopes for string theory, since it seemed to have the best possible chance of providing a fundamental unified theory. Indeed, I still think there are ideas in string theory that may be right, and its exploration has led to the uncovering of some beautiful and deep mathematics. But as a theory of the elementary particles, it has certainly so far failed, for while it initially seemed that there was only one possible consistent string theory, we now know there are a great many such theories, each apparently as consistent as the others and all leading to different universes. Thus, string theory hasn't solved the problem of how the world chooses to have the particular collection of particles and forces it does. And whatever the theory's future, I've come to doubt that it ever will.

This crisis led me to wonder whether the search for the principles that determine which laws of nature govern our world could succeed, if we continue to look only at mechanisms that act on very small scales. Instead, I began to ask myself whether there might be mechanisms that could in some way couple the properties of the elementary particles to the properties of the universe created by their inter-actions—perhaps even on astronomical and cosmological scales. By this I mean nothing mystical. Since the universe has a history, and did apparently pass through a stage when it was very small, there might be some mechanism that coupled the properties of things on the largest scales to the properties of things on the smallest scales. Thus, about five years ago I began to wonder whether there might be some way in which the properties of the elementary particles are chosen by the universe itself, during its evolution. Wondering about this made me notice and take seriously what many people had pointed out previously—that the properties of the elementary parti-cles and the conditions of the universe seem very well chosen for the universe to develop structure and life. It does seem that this is true—that if almost any other set of forces and particles had been chosen, the universe would not only not contain life, it would be much less rich in structure and variety of phenomena than our world is.

Many of the people who've noticed this have become advocates of the anthropic principle. This is the idea that the properties of the world have somehow been chosen because of—or at least are ex-plained by—the fact that with this choice intelligent life like us can exist. I'd always resisted this idea, and I still do. The anthropic prin-ciple is said to come in two forms, a weak form and a strong form. In its weak form, I think it's just the observation that the world in which we find ourselves is very special. This doesn't explain anything, it only points out the need for an explanation of *how* the world got to be special—an explanation that must be made in terms of some mechanism acting in its past. The strong form—that the laws of physics are somehow chosen in order that life can exist—is, to me, really more religion than science. Indeed, I'm not surprised to find that several advocates of the strong form of the anthropic principle are writing books and papers connecting their belief in the anthropic principle with Christian theology. This is fine, for religion, but it isn't science. Instead, when I realized that people like Martin Rees and Bernard Carr were right—that the world is very special in ways that seem a priori extremely unlikely—I began to wonder whether there might be some real mechanism, something taking place earlier in the history of the universe, that might explain how the properties of the elementary particles have been selected so that the world has the enormous amount of structure and variety it does.

At this time, I was reading a lot of biology: Richard Dawkins on evolution, Harold Morowitz on self-organization, and James Lovelock and Lynn Margulis on the Gaia idea. And I remember wondering whether, if the earth can be understood as a self-organized system, maybe the same thing was true for larger systems, such as a galaxy or the universe as a whole. This was also summertime, and I was sailing a lot, and I spent a lot of time letting the boat drift and wondering what kind of mechanisms of self-organization might have acted early in the history of the universe to select the properties of the elementary particles and forces in nature. It seemed to me that the only principle powerful enough to explain the high degree of organization of our universe—compared to a universe with the particles and forces chosen randomly—was natural selection itself. The question then became: Could there be any mechanism by which natural selection could work on the scale of the whole universe?

Once I asked the question, an answer appeared very quickly: the properties of the particles and the forces are selected to maximize the number of black holes the universe produces. This idea came right away, because of two ideas I was familiar with from my work on quantum gravity. The first is that inside a black hole, quantum effects remove the singularity that general relativity says is there—and that we know is there from the theorems of Penrose and Hawking—and a new region of the universe begins to expand as if from a big bang, there inside the black hole. I remember Bryce DeWitt, who is one of the great pioneers of quantum gravity, telling me about this idea shortly after I began to work for him, on my first postdoc. The second idea—which comes from John A. Wheeler, another great pioneer of the field—is that at such events the properties of the elementary particles and forces might change randomly. All I then needed to make a mechanism for natural selection was to assume that these changes are small, because reading Dawkins had taught me the importance for natural selection of incremental change by the accumulation of small changes in the gene. Then, with the universes as animals and the properties of the elementary particles as genes, I had a mechanism by which natural selection would act to produce universes with whatever choices of parameters would lead to the most production of black holes, since a black hole is the means by which a universe reproduces—that is, spawns another.

This was in 1989. I still don't know if the idea is right. But what

I'm very proud of is that the idea is testable. Most ideas about why the elementary particles have the properties they do which have been proposed in the past few years aren't testable. This is the main reason the field is in such a crisis. But this idea leads to a prediction, which is that if I could change any of the properties of the elementary particles the result should be either to decrease or to leave alone the number of black holes the universe makes. This is because the idea implies that almost every universe, and therefore most likely our own, has parameters that maximize the numbers of black holes it can make.

When this idea first came to me, I didn't take its prospects very seriously, and I imagine neither did most of my colleagues. I also didn't know much astrophysics, and I imagined that it would be an easy matter to test what would happen to the rate of production of black holes if you changed, for example, the mass of one or another sort of elementary particle, or the strength of one of the forces. So to test the idea, I started to learn some astronomy and astrophysics. So far, I haven't found a way to change the properties of the particles and forces to make a universe that makes more black holes, and I *have* found several changes that decrease their number. I've also brought the question to a number of astrophysicists, who know the field much better than I do. I've been very pleased that these people, some of whom I admire very much, were interested enough to spend the time to examine such an unusual idea. They made some interesting suggestions, and although no one was able to propose a change of parameters that clearly leads to the production of more black holes, several interesting possibilities, which I'm studying now, did emerge from these conversations. Certainly, if the idea's wrong, I'll be grateful if someone proposes a test that would kill it. I believe more in the general idea that there must be mechanisms of self-organization involved in the selection of the parameters of the laws of nature than I do in this particular mechanism, which is only the first one I was able to invent. But it seems that the situation at present is that there's much more testing that needs to be done, and lately I've been spending more time on this. Perhaps what's most amazing to me is that after five years this rather improbable idea is still not dead.

Whether it dies or not, I've learned enough astronomy to discover something that's completely changed my view of cosmology. This is that the idea that there are principles of self-organization acting on astronomical scales seems really to be true. During the last ten years or so, people who study galaxies have discovered evidence that feedback effects and mechanisms of self-organization are indeed happening at the level of the galaxies; they are, in fact, essential for galaxies to form stars. They're also necessary to the existence of spiral galaxies. The idea that a galaxy is a self-organized system—more an ecology than a nonliving clump of stars and gas—has become common among astronomers and physicists who study galaxies.

Thus, it seems to me quite likely that the concept of self-organization and complexity will more and more play a role in astronomy and cosmology. I suspect that as astronomers become more familiar with these ideas, and as those who study complexity take time to think seriously about such cosmological puzzles as galaxy structure and formation, a new kind of astrophysical theory will develop, in which the universe will be seen as a network of self-organized systems.

Beyond this, I also think that—whatever the fate of my ideas—this merging of the science of the fundamental and the science of the organized will overturn the usual ways of thinking about the elementary particles, too. Many of the people who work on complexity, such as Murray Gell-Mann, Stuart Kauffman, Harold Morowitz, and others, imagine that the world consists of highly organized and complex systems but that the fundamental laws are simply fixed beforehand, by God or by mathematics. I used to believe this, but I no longer do. More and more, what I believe must be true is that there are mechanisms of self-organization extending from the largest scales to the smallest, and that they explain both the properties of the elementary particles and the history and structure of the whole universe.

To put it most simply, I think a successful theory that merges relativity and cosmology with quantum theory must also be a theory of self-organization. In fact, I have an argument for this conclusion, which is based on the idea that, as Bohr taught us, quantum theory doesn't make sense unless there are clocks and observers in the world. Normally, this is no problem, because the clocks and observers are outside the system being studied, so we can just assume their existence. But if we're going to apply quantum theory to the whole universe, then there's no room for observers or clocks outside the system, because there's no "outside."

But only a complex universe—a universe complex enough to give rise to life—can have things like clocks and observers in it. And if the quantum theory of gravity requires these to exist, and if they are to exist inside the universe the theory describes, then perforce that universe must be complex, and the theory must explain why it's complex. This means there must be some relationship between quantum theory and relativity and self-organization, so that it's logically impossible to describe a relativistic, quantum-mechanical world unless mechanisms of self-organization act in that world to produce the complexity the theory needs if it's to be logically consistent.

A similar argument follows from the way space is described in Einstein's theory of general relativity. For if, as is the case, the only meaningful things in this theory are relationships between real things, then it doesn't make sense to talk about space as being made up of different points, or time as being made up of distinct moments, unless the points and the moments can be distinguished by what's happening there. This means that if it's meaningful in general relativity to speak of the world as having three continuous dimensions of space and one of time, it must be true that the view of the universe from each point of space and time is unique. Otherwise they can't be distinguished from one another. But this means the world must be complex enough so that one can tell where one is in the universe just by looking around. And, again, if the theory of general relativity requires this complexity for its consistency, it must somehow produce it, if it's to be a complete theory of the whole universe.

Thus, I believe that the question of why the laws of physics are chosen so that the world is so complex is intimately related to the basic questions about what space and time are, which we struggle with in quantum gravity. Because of this intimate relationship, I think the next years in elementary-particle physics and cosmology will be very exciting. And what's most encouraging to me is that while many of my colleagues are still depressed over string theory, some of the theoretical physicists whose imaginations I most admire—Alexander Polyakov and Holgar Nielsen, for example—are beginning to look for mechanisms by which the universe could tune the properties of its elementary particles.

Perhaps I might say a word about what it's like to work as a theoretical physicist, because it seems so different from the image I had of it when I was just beginning to dream about it. I don't know if other fields are like this, but what it feels like to work on quantum gravity is that we are on a great search, which is partly one's own search and partly one's participation in a great tradition that's also a wonderful community. Science is a very social activity; we're often traveling, and we spend enormous amounts of time talking with other people—both the friends we work with and people from the larger community. Physics is very verbal. Some of us read one another's papers—I do—but the most important channel of communication is certainly talking. In quantum gravity, there's a community of—I don't know—perhaps a few hundred people who are actively working on the problem and who are in constant communication with one another. Indeed, there's only one thing I don't like about the community of my colleagues, which is that there are still so few women. Of course, there are slowly getting to be more women in the field, but this isn't happening as fast as it is in other fields. It's a very interesting question to ask why this is.

There's another side to doing fundamental science which isn't at all social: it's one's personal confrontation with nature. In the end, I'm trying to understand things like the meaning of time because of my need to know who I am, what this world is, what I'm doing here. To do science is, for me, one kind of response to the alienation of being a small creature in an enormous world. Part of being a scientist, for me, is that in the end I know that I alone am responsible for what I believe.

As a scientist, one can believe what one wants and work on what one wants, but one also accepts the idea that in the end the community is the ultimate judge of the usefulness of what one does. This requires an ethics that makes honesty and respect for the views of others essential, while at the same time making individuality and difference and disagreement essential. So at any one time in the scientific community, there's a consensus about certain matters on which almost everyone has come, after long struggle, to agree; but there's also a large area where no consensus exists. Indeed, this state of affairs is necessary, because if there was too great a consensus the process would stop; this would be the death of science.

MARTIN REES: One of the key issues in physics is to reconcile gravity with the quantum principle and the microphysical forces. There are various schools of thought; the Stephen Hawking School, the Roger

Penrose School, and a number of others. My view is that we're a long way away from a consensus in that field, but Smolin and Ashtekar have injected important new ideas into that debate.

Quantum gravity was one of the subjects beyond the fringe, when John Wheeler talked about it in the 1950s. Now it's something where serious approaches are being adopted. But we're still a long way from experimental test. Lee Smolin's most important insight was to suggest a new way of looking at space and time in terms of a lattice structure on a tiny scale. It relates in a way to Wheeler's very farsighted ideas of spacetime foam: the idea that if you look at space and time on a very tiny scale, there are no longer three dimensions of space and one of time but the dimensions all get screwed up in a complicated way.

The other idea with which Smolin is associated is "natural selection" of universes. He's saying that in some sense the universes that allow complexity and evolution reproduce themselves more efficiently than other universes. The ensemble itself is thus evolving in some complicated way. When stars die, they sometimes form black holes. (This is something which I wear my astrophysical hat to study.) Smolin speculates—as others, like Alan Guth, have also done—that inside a black hole it's possible for a small region to, as it were, sprout into a new universe. We don't see it, but it inflates into some new dimension. Smolin takes that idea on board, but then introduces another conjecture, which is that the laws of nature in the new universe are related to those in the previous universe. This differs from Andrei Linde's idea of a random ensemble, because Smolin supposes that the new universe retains physical laws not too different from its parent universe. What that would mean is that universes big and complex enough to allow stars to form, evolve, and die, and which can therefore produce lots of black holes, would have more progeny, because each black hole can then lead to a new universe; whereas a universe that didn't allow stars and black holes to form would have no progeny. Therefore Smolin claims that the ensemble of universes may evolve not randomly but by some Darwinian selection, in favor of the potentially complex universes.

My first response is that we have no idea about the physics at these extreme densities, so we have no idea whether the physics of the daughter universe would resemble that of the parent universe. But one nice thing about Smolin's idea, which I don't think he realized himself in his first paper, is that it's in principle testable, because we know enough about how stars evolve, and we know what stars turn into black holes and what stars turn into neutron stars. That's one of the things my colleagues and I have worked on.

We know enough to work out how the number of black holes forming would change if we tweak the laws of physics. Suppose we just changed the strength of gravity, or changed the mass of the neutron a bit. How would that change stellar evolution, and how would it change the propensity of stars to end up as black holes? If Smolin is right, and if this ensemble has been evolving through enough successive "generations" of universes, then our universe ought to have the property that it maximizes the number of progeny. It ought to be governed by laws that give it an evolutionary advantage, so that it maximizes the number of progeny it contributes to the ensemble. That's testable, because we can ask: If the laws of nature changed a bit, would that slightly altered universe make fewer black holes than ours does? If it turns out that our universe has the properties that maximize the number of black holes that form in it, that would be evidence for Smolin's being correct.

The bad news is that I don't see any reason to believe that our universe has the property that it forms more black holes than any slightly different universe. There are ways of changing the laws of physics to get *more* black holes, so in my view there are arguments *against* Smolin's hypothesis. It's just ordinary everyday physics, or *fairly* everyday physics, that determines how stars evolve and whether black holes form, and I can tell Smolin that our universe doesn't have the properties that maximize the chance of black holes. I could imagine a slightly different universe that would be even better at forming black holes. If Smolin is right, then why shouldn't our universe be like that? We may be able to disprove Smolin, so in that sense his conjecture is a genuine scientific theory in that it's refutable.

MURRAY GELL-MANN: Smolin? Oh, is he that young guy with those crazy ideas? He may not be wrong!

ROGER PENROSE: Smolin's view on the bridge between the quantum and the classical levels in physics is somewhat different from mine. I talk to him a fair amount about it. He has a very good grasp of contemporary physics, but is appropriately critical of it; he knows its limitations and has put forward interesting ideas for developing physics into something better. I've always thought of him as a very powerful critical physicist.

What Lee Smolin and Carlo Rovelli have developed with regard to the underlying structure of the universe, I find extremely interesting. Where it will ultimately go I do not know; it's certainly one of the more promising ideas that I've seen.

PAUL DAVIES: Lee Smolin I only just met. I warm to scientists who have a freewheeling mind and really pursue their ideas to the logical extreme—John Archibald Wheeler is another—without taking that extreme too seriously. Physics and cosmology are wonderlands for bizarre speculation, which serves a useful scientific purpose without having to be right, though it may be!

ALAN GUTH: Lee Smolin came into the relativity business later than Hawking or Penrose, so he had to deal with a different class of problems. His work is aimed not at classical general relativity, the way much of Penrose is and Hawking's famous work has been, but rather at quantum general relativity—that is, quantum gravity.

General relativity as formulated by Einstein was a classical theory, by which I mean all of the quantities that appear in the theory have definite values at all times, and the equations tell you how those quantities evolve in time. There are no probabilities in a classical theory like general relativity. Everything is unambiguous. However, physicists learned in the early part of the twentieth century that the real world is not quite like that.

The real world is described by quantum theory, and in a quantum theory nothing can ever be measured precisely, even in principle. There are always uncertainties about the current state of the universe, or any piece of the universe, and when you make predictions about how a given system will behave, the best predictions you can possibly make are probabilistic predictions. You predict that such and such an outcome will happen with one-third probability, and another outcome will happen with 17-percent probability, and so on. In many cases, of course, those probabilities can be very close to one; in some cases, you can tell that something will happen with 99.999-

percent probability, but it's always fundamentally a probabilistic prediction, if one is talking quantum theory.

Everybody now believes that general relativity has to be merged with quantum theory, to produce a correct description of how gravity and space behave. So far we've had only mixed success in that venture. When one tries to combine general relativity with quantum mechanics using the same approach that's been successful for combining electromagnetism with quantum mechanics, what one finds is that this approach just doesn't work. When you do the calculations, you find that many of the quantities turn out to be infinite, and nobody knows what to do about that.

We've been looking for other approaches, and that's where Lee Smolin's work has been concentrated. The majority approach, which hasn't been Lee's approach, has been from people who came from particle theory, and those people are mostly of the opinion that the solution to the problems of quantum gravity lies in superstrings. Superstrings is a completely new theory, in which you assume—for reasons that are very hard to trace but are valid reasons—that the fundamental entity in nature is a microscopic string, an object that has essentially negligible width and a very small length, and that these funny things make up the fundamental entities, of which we're seeing only the very low-energy consequences.

The basic motivation behind these superstrings is to build a quantum theory of gravity that gives finite answers. It has been shown—at least, for the kinds of calculations that people know how to do—that the problem of the infinities of gravity are avoided by these superstring theories. What the superstring people have yet to do, however, is show that the theory has anything to do with reality; that is, they have not yet been able to explain how to extract the low-energy consequences of the theory—to show that superstrings really do produce the world we see.

A possible reason that *Discover* magazine dubbed Lee "The New Einstein" on a recent cover is that his work is motivated by the same goal—to construct a unified theory of physics—and his approach is to keep Einstein's original theory as the fundamental basis of it. Superstring theory basically puts Einstein's theory in the background. The belief is that Einstein's theory will reemerge as a low-energy limit, but it's not the fundamental ingredient of the theory. The fun-

damental ingredient of the superstring theory is this microscopic
string. In Smolin's formulation, the fundamental ingredient remains
the gravitational field, and the goal is to treat it quantum mechanically. What he hopes to do that's different from the failed approach—
the approach that successfully quantizes electromagnetism but fails
for gravity—is to exploit the fact that the theory of gravity is fundamentally nonlinear.

In this case, the nonlinearity can be explained in plain physical
terms: in electromagnetism, the carrier of the interaction is the photon, the particle of light; for gravity, there's a hypothetical carrier, the
graviton, which plays the analogous role to the photon. The important difference is that photons don't produce photons. Gravitons,
however—since they carry energy, and any form of energy creates a
gravitational field—do create gravitons. It's this complication that
leads to all the other complications associated with trying to build a
quantum theory of gravity. Because gravitons can produce themselves, the entire theory becomes much, much more complicated
and leads to tremendously difficult problems, in terms of avoiding infinities that seem to arise when one tries to calculate.

The relativity physicists belong to a small club. It's a club that has
yet to convince the majority of the community that the approach
they're pursuing is the right one. Certainly Smolin is welcome to
come and give seminars, and at major conferences he and his colleagues are invited to speak. The physics community is interested in
hearing what they have to say. But the majority looks to the superstring approach to answer essentially the same questions.

Chapter 18

PAUL DAVIES

"The Synthetic Path"

ALAN GUTH: *Paul Davies is a good popularizer. He's also a good physicist. He's known mostly for his work in the area of attempts at quantum gravity, although he's not approaching exactly the same problem as either Lee Smolin or the people who do superstring theory are. He's the kind of person who takes a more pragmatic approach.*

• • •

PAUL DAVIES is a theoretical physicist; professor of natural philosophy at the University of Adelaide; author of many books, including Other Worlds (1980), God and the New Physics (1983), Superforce (1984), The Cosmic Blueprint (1989), and The Last Three Minutes (1994); coauthor with John Gribbin of The Matter Myth (1992).

PAUL DAVIES: People are interested in questions of origins. I'm referring to the origin of the universe, but the origin of life and the origin of consciousness are equally major landmarks in trying to understand what we are and how we fit into the wider scheme of things. It's interesting that for those who are religious and insist on having a role for God, there are only three gaps left in our knowledge where they would wish to invoke God as a direct influence in the world. One is the origin of consciousness—or the human soul, if you like. The second is the origin of life: life getting started from nonlife. The third is the origin of the universe as a whole. These are the three perceived gaps in science where people would wheel in God, if you like. If people aren't as fascinated by the origin of life or consciousness as they are by the origin of the universe, there's something a bit wrong about the way these subjects are being presented. From the point of view of human beings, they're equally profound and equally as important.

At quite an early age, I became closely associated with the so-called arrow-of-time problem. This has to do with the mystery of why most physical processes in the universe seem to go one way in time, whereas the underlying laws of physics that govern them are reversible—they have no preferred time orientation. I got into this because of a couple of papers by John Wheeler and Richard Feynman, in which they tried to explain how, for example, radio signals always arrive at the receiver after they leave the transmitter and never before. They had a clever way of starting out with time-symmetric electromagnetic waves (forwards and backwards in time, symmetrically), and recovering the purely time-forwards waves by appealing to cosmology—that is, taking into account a whole universe full of emitters and absorbers of electromagnetic waves. This led me to investigate a wide range of other topics in which time symmetry gets broken. When I was twenty-four, I wrote a book on the subject, called *The Physics of Time Asymmetry*. It was really just a preliminary skirmish with an enormously complicated topic, but lots of influential people, like Wheeler, Roger Penrose, and Martin Gardner, said nice things about it. Even Feynman recommended it to a colleague!

In terms of actual discoveries, my name is most often coupled to a weird effect I found in the theory of quantum fields. Imagine a total vacuum, devoid of all particles including photons. Now suppose you accelerate through that emptiness, what do you see? Nothing? In fact, you see a bath of heat radiation, even though your nonaccelerating friend still sees absolutely nothing. The effect, which is closely related to Stephen Hawking's discovery that black holes radiate heat, was discovered independently by Bill Unruh, at the University of British Columbia. I wrote up this result in the mid-seventies, almost casually. The effect is very small, and not hard to prove, and I didn't think many people would be interested, but papers still appear at the rate of several a year, elaborating on this or that aspect of "acceleration radiation."

After this success, I worked on the theory of quantum fields in curved spacetime—that is, in the presence of gravitational fields. The book I wrote with my student Nick Birrell, called *Quantum Fields in Curved Space*, remains the principal text on the subject, I am pleased to say. A lot of my investigations concerned the behavior of quantum fields in certain model universes that were simple to study. We were interested in many questions. Could the expansion of the universe create particles? How did the quantum field get disturbed by the gravitational field of the universe, and how did this disturbance in turn react gravitationally? One of these model universes is named after the Dutch cosmologist Willem de Sitter, and together with another student, Tim Bunch, I spent a lot of time looking at it. Among many results emerged the concept of a particularly interesting quantum-vacuum state, still known as the Bunch-Davies vacuum. It never occurred to me at the time—the late seventies—that this stuff would find any real applications. How delighted I was when suddenly de Sitter space became of central importance in Alan Guth's inflationary-universe scenario, and people began using the Bunch-Davies vacuum in their calculations!

I also did a lot of work on black holes and their thermodynamic properties, discovering, for example, that if a black hole carries a large enough electric charge it can remain in equilibrium with a surrounding heat bath instead of evaporating away in the manner Hawking first described. I've always been fascinated by Penrose's belief that gravitation represents a sort of entropy in its own right, and many papers I wrote in the eighties were attempts to flesh out this idea, but without complete success.

It's very curious how Alan Guth got into the inflationary scenario. He was trying to solve a rather specific problem connected with magnetic monopoles. The standard hot-big-bang theory, combined with our best knowledge of particle physics, indicated that the universe ought to be stuffed full of magnetic monopoles, and yet we don't see any. The question was, How had they been eliminated? One obvious way to get rid of them is to have the universe inflate by some large factor that simply dilutes the density of these monopoles.

Guth wasn't a cosmologist; he was a particle physicist trying to get rid of monopoles, and so he proposed the idea that the universe, during its first split second, suddenly jumped in size by a huge amount. Plausible answers to key cosmological questions—such as whether the universe is expanding at precisely the rate to escape its own gravitational pull, and whether the quantum fluctuations around that precise rate would give the sort of spectrum just observed by the COBE satellite—came as a bonus. The fascinating thing is that what Guth did was to come in the back door and discover this immensely rich seam of ideas, which he then successfully quarried. His inflationary theory, inevitably refined, is now pretty much the standard cosmological scenario for the origin of the universe.

Only twenty-five years ago it was not considered appropriate to consider the physical mechanism of the birth of the universe. I remember a lecture I attended as a graduate student at University College London. This was a couple of years after the discovery in 1965 of the cosmic microwave background radiation, and the implications of that discovery had not yet generally sunk in. A professor was talking about how theorists had computed, based on the existence of this radiation, that there would be about 25 percent helium and 75 percent hydrogen in the universe, and that this had come from an analysis of the nuclear processes that took place in the first few minutes after the big bang. Everyone in the lecture hall fell about laughing, because they thought it was so absurd and audacious to talk about the first three minutes after the big bang, just on the basis of the discovery of this radiation. Now, of course, it is absolutely standard cosmological theory. We feel we understand the first few minutes of the universe very well.

What we now find is that the big bang has gone from being merely a description of the origin of the universe to being an explanation. That's a key difference. Simply saying that things just happen that way—in other words, to say that things are the way they are because they were the way they were—constitutes a description. What we now have is something much closer to a scientific explanation, in which not only can we account for the fact that there was a bang but also a lot of the specific features of the big bang now emerge from well-formulated physical theory, instead of being put in as ad-hoc initial conditions. That's the big difference. The latest COBE discovery adds enormously to the strength of the big-bang theory as a proper theory and not just a description.

The evolution of the big-bang theory leads to a discussion of the anthropic principle, which says that the world we see must reflect, to some extent, the fact that we're here to see it—not only here but here at this particular location in space and time. There are different variants of the anthropic cosmological principle, and how much credence you can give it depends on which one you're talking about. What's quite clear is that there must be an anthropic companion to our science. To take a trivial and extreme example: most of the universe is empty space, and yet we find ourselves on the surface of a planet. We're therefore in a very atypical location, but of course it's no surprise that we're in this atypical location, because we couldn't live out there in space.

Obviously, there's an anthropic factor to what we observe and the position in the universe from which we observe it, or maybe the time, the epoch, that we observe it. Having said that, the question is whether it's just a comment about the universe or in some sense an explanation for some features of the universe. If there's only one universe, it's just a comment on it. But if we imagine that there is a whole ensemble of universes—a huge variety, with different conditions, different laws—then it starts to become an explanation, or a selection principle. Part of the reason for the order we observe in the universe is that this is one of the few universes out of the whole ensemble that is cognizable. Some people have tried to carry this principle to a ludicrous extreme by making out that ultimately there are no laws of nature at all, that there is only chaos, that the lawfulness of the universe is merely explained by the fact that we've selected it from this infinite variety of essentially chaotic worlds. That is demonstrably false, and an unreasonable extrapolation of the whole anthropic idea.

It's remarkable that the universe is lawful, that there exist underlying rational principles which govern the way the universe behaves. We can't account for that just on the basis of the fact that we're here to see it, as some people have tried to do. There's a dual principle at work. There's a principle of rationality that says that the world is fashioned in a way that provides it with a rational order, a mathematical order. There's a selective principle—which is an anthropic principle—that says that maybe out of a large variety of different possible worlds this type of world is the one we observe.

We can't avoid some anthropic component in our science, which is interesting, because after three hundred years we finally realize that we do matter. Our vantage point in the universe is relevant to our science. But it's very easy to misconstrue the anthropic principle, and draw ridiculous conclusions from it. You have to be very careful how you state it. What it is *not* saying is that our existence somehow exercises a theological or causative compulsion for the universe to have certain laws or certain initial conditions. It doesn't work like that. We're not, by our own existence, creating such a universe.

We are now very close to identifying the nature of the fundamental building blocks out of which the world is put together. This reductionist path is tremendously important and has exercised an enormous influence in the thinking of physicists, but it's only a part of the story. To say that the world is built up of a collection of certain particles playing certain roles of interaction is one thing. But to give an explanation of problems like the origin of life, the origin of consciousness—problems that refer to highly complex systems—that's quite another. To talk about complexity, we have to realize that there are systems the behavior of which can be understood only by looking at the collective and organizational aspects, instead of the individual constituents. It's impossible to explain the behavior of these so-called adaptive complex systems in a purely reductionistic manner and expect to build up from that. These are systems like biological organisms, which appear to respond to and adapt in accordance with their environment.

I look forward to a time when the biologists stop berating the physicists for abandoning reductionism. At the moment, the biologists are strongly and evangelically reductionistic, and any suggestion by physicists that one can deviate from the path of strict reductionism tends to evoke a rap over the knuckles from the biologists. My personal belief is that biologists tend to be uncompromising and reductionistic because they're still feeling somewhat insecure with their basic dogma, whereas physicists have three hundred years of secure foundation for their subject, so they can afford to be a bit more freewheeling in their speculation about these complex systems. I hope to see this cultural division between these two communities dissolve away within the next ten to twenty years, so that they'll be able to talk to each other in the same language.

There are two paths in investigating the world: the reductionist path and the synthetic path. In the science of complexity, it's essential to recognize that there is this second path. Complexity amounts to more than mere complication. It's more than just a large number of simple systems coming together in conjunction. Complex systems really do have their own laws and principles, and their own internal logic.

In the next few decades, physics will be going in the direction of complexity. One of the key questions for physics is, Can the reductionist program be completed? Stephen Hawking said, in his famous 1979 address on his inauguration to the Lucasian Chair, that the end might be in sight for theoretical physics, by which he meant that the end of this reductionist program might be in sight. Indeed, we may complete it and be able to write down a formula you could wear on your T-shirt—some mathematical statement, or a set of principles encapsulated in a single piece of mathematics, describing all the fundamental particles and forces out of which the world is built.

That would still leave this path of complexity, this synthetic or holistic way of looking at the world. There, what I see as the real excitement is the dissolving away of the division between physics and biology. We see a very curious phenomenon at the moment: while physicists are increasingly recognizing the importance of looking at the collective, organizational, and qualitative features of complex

systems, and recognizing that they have their own laws and principles and qualities, in a way that makes them every bit as fundamental as the elementary particles out of which the world is built, at that
same time the biologists are going the other way, becoming overly reductionistic and regarding life as nothing but a collection of individual particles interacting in an unwitting manner by means of blind
and purposeless forces.

It's often said that if we had a theory of everything, everything
would be explained. But when physicists talk about a theory of everything, they don't mean literally everything. They don't mean a theory
that would explain how the stock market rises and falls, still less
something that would explain the origin of life. They mean a theory
that accounts for all these fundamental units out of which the world
is built.

MARTIN REES: I've known Paul Davies since he was a postdoc at the
Institute of Astronomy, at Cambridge; he's just a bit younger than I
am. Back then, he wrote his first book, *The Physics of Time Asymmetry*. Since that time, he's gone from strength to strength as an exposi-
tor of physics. His books are remarkably comprehensible and clear,
and deserve their success. Heinz Pagels is the only person I'd put
above him. But I'm not sure I'd put any of the other writers on cosmology or particle physics on the level of Paul Davies, in terms of
clarity and fairmindedness.

ALAN GUTH: Paul Davies is a good popularizer. He's also a good physicist. He's known mostly for his work in the area of attempts at quantum gravity, although he's not approaching exactly the same problem
as either Lee Smolin or the people who do superstring theory are.
He's the kind of person who takes a more pragmatic approach. By
that I mean there are subsidiary problems—ways of approaching it
where the goal is to solve half the problem rather than the whole
problem, and that's Davies' approach; while the approach of Smolin,
if it's right, is a solution to the whole problem, and the same applies
to the superstring people.

Davies is known for doing work on quantum field theory in a
curved-space background. What that means is that he's solving half
the problem, by treating the matter fields that describe electrons,
protons, neutrons, and photons (photons count as matter, in this

context) fully relativistically and quantum mechanically, but at the same time treating gravity classically. That turns out to be a well-defined but difficult problem. It's a problem that doesn't seem to have fundamental difficulties, but in practice it has many difficulties, and many of the important calculations were done first by Paul Davies.

Part Four

WHAT WAS DARWIN'S
ALGORITHM?

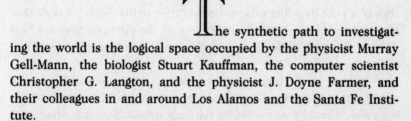

The synthetic path to investigating the world is the logical space occupied by the physicist Murray Gell-Mann, the biologist Stuart Kauffman, the computer scientist Christopher G. Langton, and the physicist J. Doyne Farmer, and their colleagues in and around Los Alamos and the Santa Fe Institute.

The Santa Fe Institute was founded in 1984 by a group that included Gell-Mann, then at the California Institute of Technology, and the Los Alamos chemist George Cowan. Some say it came into being as a haven for bored physicists. Indeed, the end of the reductionist program in physics may well be an epistemological demise, in which the ultimate question is neither asked nor answered but instead the terms of the inquiry are transformed. This is what is happening in Santa Fe.

Murray Gell-Mann, widely acknowledged as one of the greatest particle physicists of the century (another being his late Caltech colleague, Richard Feynman), received a Nobel Prize for work in the 1950s and 1960s leading up to his proposal of the quark model. At a late stage in his career, he has turned to the study of complex adaptive systems.

Gell-Mann's model of the world is based on information; he connects the reductionist, fundamental laws of physics—the simple rules—with the complexity that emerges from those rules and with what he terms "frozen accidents"—that is, historical happenstance. He has given a name to this activity: "plectics," which is the study of simplicity and complexity as it is manifested not just in nature but in such phenomena as language and economics. At the institute, he provides encouragement, experience, prestige, and his vast reservoir of scientific knowledge to a younger group of colleagues, who are mostly involved in developing computational models based on simple rules that allow the emergence of complex behavior.

Stuart Kauffman is a theoretical biologist who studies the origin of life and the origins of molecular organization. Twenty-five years ago, he developed the Kauffman models, which are random networks exhibiting a kind of self-organization that he terms "order for free." Kauffman is not easy. His models are rigorous, mathematical, and, to many of his colleagues, somewhat difficult to understand. A key to his worldview is the notion that convergent rather than divergent flow plays the deciding role in the evolution of life. With his colleague Christopher G. Langton, he believes that the complex systems best able to adapt are those poised on the border between chaos and disorder.

Kauffman asks a question that goes beyond those asked by other evolutionary theorists: if selection is operating all the time, how do we build a theory that combines self-organization (order for free) and selection? The answer lies in a "new" biology, somewhat similar to that proposed by Brian Goodwin, in which natural selection is married to structuralism.

Christopher G. Langton has spent years studying evolution through the prism of computer programs. His work has focused on abstracting evolution from that upon which it acts. He has created "nature" in the computer, and his work has given rise to a new discipline called AL, or artificial life. This is the study of "virtual ecosystems," in which populations of simplified "animals" interact, reproduce, and evolve. Langton takes a bottom-up approach to the study of life, intelligence, and consciousness which resonates with the work of Marvin Minsky, Roger Schank, and Daniel C. Dennett. By vitalizing abstraction, Langton hopes to illuminate things about life that are not apparent in looking at life itself.

J. Doyne Farmer is one of the pioneers of what has come to be called chaos theory—the theory that explains why much of nature appears random even though it follows deterministic physical laws. It also shows how some random-seeming systems may have underlying order which makes them more predictable. He has explored the practical consequences of this, showing how the game of roulette can be beaten using physics; he has also started a company to beat the market by finding patterns in financial data.

Farmer was an Oppenheimer Fellow at the Center for Nonlinear Studies at the Los Alamos National Laboratory, and later started the complex systems group, which came to include some of the rising stars in the field, such as Chris Langton, Walter Fontana, and Steen Rasmussen. In addition to his work on chaos, he has made important theoretical contributions to other problems in complex systems, including machine learning, a model for the immune system, and the origin of life.

Chapter 19

MURRAY GELL-MANN
"Plectics"

J. DOYNE FARMER: *The first thing that makes me respect Murray
is that unlike all his contemporaries, including Feynman,
Weinberg, Hawking, and all the other particle physicists, he
saw that complexity is the next big problem. The kind of
breakthroughs he made in the early 1960s in terms of impact
on the world of science are not going to get made in that domain, they are going to get made in this domain. Murray recognized that, and has become more than just conversant with
what's going on and with what the problems are.*

• • •

MURRAY GELL-MANN is a theoretical physicist; Robert
Andrews Millikan Professor Emeritus of Theoretical
Physics at the California Institute of Technology; winner
of the 1969 Nobel Prize in physics; a cofounder of the
Santa Fe Institute, where he is a professor and cochair-
man of the science board; a director of the J.D. and C.T.
MacArthur Foundation; one of the Global Five Hundred
honored by the U.N. Environment Program; a member of
the President's Committee of Advisors on Science and
Technology; author of The Quark and the Jaguar: Adven-
tures in the Simple and the Complex *(1994).*

MURRAY GELL-MANN: When I was a small child, I was very interested in natural history and linguistics and archaeology. Though I lived in New York City, I managed to find some patches of country where I could become familiar with birds and butterflies and trees and flowering herbs. Even then, I was fascinated by the results of biological evolution and of the evolution of human culture. So it's not unnatural that I would want to try to understand the chain of relationships linking the fundamental physical laws that govern all matter in the universe to the behavior of the rich complex fabric we see around us and of which we are a part.

One way to make the task manageable is to look at the world from the point of view of information. When we do that, we see that the basic pattern is one of complexity emerging from very simple rules, initial order, and the operation, over and over again, of chance. In the case of the whole universe, the fundamental laws of physics constitute those simple rules.

There are various quantities labeled "complexity." In each case, the complexity of a thing is context-dependent—in other words, dependent not only on the thing being described but also on who or what is doing the describing. There's one quantity in particular that I think most deserves the label—what I call "effective complexity." A related quantity, which I have named "potential complexity," is also very important. Neither is yet defined with mathematical rigor, and that's a task I've undertaken. Some of the other quantities that people have called "complexity" are also well worth discussing.

In any case, to refer to the subject on which some of us now work as "complexity" seems to me to distort the nature of what we do, because the simplicity of the underlying rules is a critical feature of the whole enterprise. Therefore what I like to say is that the subject consists of the study of simplicity, complexity of various kinds, and complex adaptive systems, with some consideration of complex nonadaptive systems as well. To describe the whole field, I've coined the word "plectics," which comes from the Greek word meaning "twisted" or "braided." The cognate Latin word, *plexus*, also meaning "braided," gives rise to "complex," originally "braided together." The related Latin verb *plicare*, meaning "to fold," is connected with *simplex*, originally "once-folded," which gives rise to "simple."

Plectics is then the study of simplicity and complexity. It includes the various attempts to define complexity; the study of the roles of simplicity and complexity and of classical and quantum information in the history of the universe; the physics of information; the study of nonlinear dynamics, including chaos theory, strange attractors, and self-similarity in complex nonadaptive systems in physical science; and the study of complex adaptive systems, including prebiotic chemical evolution, biological evolution, the behavior of individual organisms, the functioning of ecosystems, the operation of mammalian immune systems, learning and thinking, the evolution of human languages, the rise and fall of human cultures, the behavior of markets, and the operation of computers that are designed or programmed to evolve strategies—say, for playing games or solving problems.

The Santa Fe Institute, which I helped to found in 1984, gathers together mathematicians, computer scientists, physicists, chemists, neurobiologists, immunologists, evolutionary biologists, ecologists, archaeologists, linguists, economists, political scientists, and historians, among others. The emphasis is on interactive people. Many distinguished scientists and scholars yearn to stray outside their own fields but can't do so easily at their own institutions. We didn't want to locate our institute near Harvard or Stanford, where there's enormous pressure of received ideas—ideas accepted by a whole community and therefore difficult to challenge. In Santa Fe, we can think and talk freely, constrained only by the need to agree with reality.

The poet Arthur Sze wrote, "The world of the quark has everything to do with a jaguar circling in the night." What is the key to understanding the jaguar circling in the night, from the point of view of information? The major insight here is that perceived regularities in the stream of data reaching a complex adaptive system—one that can adapt, learn, or evolve the way living things on Earth evolve—are compressed into models or schemata. Those schemata are subject to change and to replacement by other schemata, so that various alternative schemata compete. When the schemata are used to describe or predict the behavior of the world or to prescribe behavior for the complex adaptive system itself, there are real-world consequences. Those consequences feed back to influence the competition among schemata, and that's how learning and adaptation take place.

The theory of complex adaptive systems, which we're now beginning to develop, should apply to all such systems, wherever they occur in the universe. Just think how many galaxies there are in the universe and how many stars there are in each galaxy. Many of those stars presumably have planets that can support complex adaptive systems. We don't know yet what constraints physical laws impose on the nature of such systems. Must they resemble, to some extent, life on Earth or machines constructed by living organisms on Earth? Or can they take very different forms? We don't know, for example, whether biochemistry on Earth is nearly unique or whether it was just one of many possibilities. In other words, we're not yet sure to what extent biochemistry was determined by physics and to what extent it was determined by the accidents of history.

I mentioned that the effective complexity of the world around us comes from very simple rules and initial order, plus the operation of chance, which is associated with indeterminacy. The most fundamental source of indeterminacy is quantum mechanics, the basic framework for physical law. In contrast to the older classical physics, quantum mechanics is not fully deterministic. Even if the initial condition of the universe and the fundamental law of the elementary particles and their interactions are both exactly known, the history of the universe is still not determined. Instead, quantum mechanics gives only probabilities for alternative histories of the universe. In some situations, those probabilities are nearly certainties, and classical physics is a good approximation, but in other situations the indeterminacy is striking. For example, when a radioactive nucleus disintegrates, emitting an alpha particle, say, the direction of emission of that particle is altogether unknowable in principle before the disintegration takes place—all directions are equally probable.

Even in the classical approximation, with the fundamental law assumed to be exactly known, effective indeterminacy of the future still arises from partial ignorance of present circumstances (which are actually in part the results of earlier accidents) and from difficulty of calculation. This kind of indeterminacy is exacerbated by the common phenomenon of chaos in nonlinear systems, which refers to an extreme sensitivity of the outcome to details of the present situation.

The importance of accidents in the history of the universe can thus hardly be exaggerated. Each of us human beings, for example, is the product of an enormously long sequence of accidents, any of which could have turned out differently. Think of the fluctuations that produced our galaxy, the accidents that led to the formation of the solar system, including the condensation of dust and gas that produced Earth, the accidents that helped to determine the particular way that life began to evolve on Earth, and the accidents that contributed to the evolution of particular species with particular characteristics, including the special features of the human species. Each of us individuals has genes that result from a long sequence of accidental mutations and chance matings, as well as natural selection.

Now, most single accidents make very little difference to the future, but others may have widespread ramifications, many diverse consequences all traceable to one chance event that could have turned out differently. Those we call frozen accidents. I give as an example the right-handed character of some of the molecules that play important roles in all life on Earth though the corresponding left-handed ones do not. People tried for a long time to explain this phenomenon by invoking the left-handedness of the weak interaction for matter as opposed to antimatter, but they concluded that such an explanation wouldn't work. Let's suppose that this conclusion is correct and that the right-handedness of the biological molecules is purely an accident. Then the ancestral organism from which all life on this planet is descended happened to have right-handed molecules, and life could perfectly well have come out the other way, with left-handed molecules playing the important roles.

Another example can be chosen from human history. For instance, Henry VIII became king of England because his older brother Arthur died. From the accident of that death flowed all the coins, all the charters, all the other records, all the history books mentioning Henry VIII; all the different events of his reign, including the manner of separation of the Church of England from the Roman Catholic Church; and of course the whole succession of subsequent monarchs of England and of Great Britain, to say nothing of the antics of Charles and Diana. The accumulation of frozen accidents is what gives the world its effective complexity.

The effective complexity of something is the length of a brief description of its regularities. Those regularities can come from only two sources: the fundamental laws, which are very simple and briefly describable, and frozen accidents.

As time goes on, systems of greater and greater effective complexity appear. That's true for nonadaptive systems, such as galaxies, stars, and planets, as well as for complex adaptive systems, as in biological evolution. Of course, I don't mean that each individual system becomes more complex. Some things get simpler; they may even disappear altogether, as in the case of vanished civilizations. Instead of a steady march toward greater complexity everywhere, there's a tendency for the envelope of effective complexity to expand. We can understand why. With the passage of time, more and more accidents occur, and frozen accidents accumulate. In fact, at any time, there are many mechanisms at work producing self-organization, which results in local order, even though the average disorder in the universe is increasing in accordance with the second law of thermodynamics. Self-organization gives rise, for example, to the arms of spiral galaxies and the myriad symmetrical shapes of snowflakes.

In the case of complex adaptive systems, their schemata have consequences in the real world, which exert selection pressures back on the competition among the schemata, and those schemata that produce favorable results in the real world have a tendency to survive, or to be promoted, and those that are less successful in the real world have a tendency to be demoted or to disappear. In many situations, complexity may offer a selective advantage. It is a challenge to evolutionary biologists, for example, to understand when that is the case.

Light can be thrown on many such questions by making use of computer-based complex adaptive systems, which can be used (1) to provide crude models of natural complex adaptive systems, (2) to supply interesting examples of complex adaptive systems for study, (3) to evolve new strategies for playing games or for solving problems, or (4) to solve problems by means of "adaptive computation."

The study of computer-based complex adaptive systems is already burgeoning, especially as a mathematical discipline concerned with the relation between simple rules and the emergence of complex behavior. That's something worth pursuing in its own right, but even more exciting is the possibility of useful contributions to the life sciences, the social and behavioral sciences, and even matters of policy for human society.

The favorite activity of some of my colleagues, especially my younger colleagues, at the Santa Fe Institute and of their friends around the world is to construct computer models with very, very simple rules—carefully chosen, stripped-down sets of rules that permit complex behavior to emerge. It's a remarkable and somewhat addictive experience to watch that emergence. We have people who are very good at stripping down rules for computer models—the political scientist Bob Axelrod, for example. He also has a flair for persuading his colleagues in political science that such a simplified model is somehow relevant to reality. If I came up with a model of that kind and presented it in a lecture to political scientists, they'd laugh me off the platform. Bob, however, presents it in such a way that social scientists can accept it. For example, imagine a circle of little polities occupying the coast of a Polynesian island with a huge volcano in the middle. The polities interact with one another either by forming alliances or making war. Each one can attack only an immediate neighbor or one that can be reached through an uninterrupted sequence of allies. Somehow Axelrod manages to extract interesting lessons from such a trivial, one-dimensional model.

Someday we'll have a full-fledged mathematical science, with theorems and proofs, that will make it clear, for instance, when new rules merely complicate the picture without adding anything essential to the emergent patterns. The construction of that science lies at one end of the spectrum of efforts to use computers to help us think about complicated systems. At the other end of the spectrum are attempts to think about policy problems that humanity faces in the real world, in connection with human society, the rest of the biosphere, and the relation between the two. In the middle, we have attempts to understand better the operation of complex adaptive systems in the life sciences and in the behavioral and social sciences. When we get away from the mathematical end of the spectrum, the accumulation of accidents of history enters in a very important way.

The stripped-down computer models are typically ones that apply, in a general way, to complex adaptive systems on any planet in the universe. They don't contain any historical information about the planet Earth, or about the organisms that inhabit the planet Earth, or about human beings and the institutions we've built.

In the simple exercises that are so popular, one starts with a caricature of one level of organization, and then one often sees a higher level of organization emerge. Starting with highly simplified individuals, you may see the emergence of a society. Starting with highly simplified polities, you may see confederations emerge. Suppose, however, you want a simplified description of human society as it exists on this planet, with all its polities and the various levels—federations, confederations, and so on—that exist, and their various relations with one another, the results of a huge number of historical accidents. These entities are all historical and peculiar to this planet and to human beings. You're forced to start complicating the stripped-down models by adding in other things—especially, new levels of organization—without waiting for them to emerge. You don't wait for the individuals in your model to develop a city or a business firm, and you don't depend on the cities and the firms to invent a nation, and the nations to invent a U.N. You have to put a lot of that in, along with some of the special properties that human beings and their firms, cities, ethnic groups, nations, and international organizations exhibit on this planet. You can no longer be content with the thrill that my friends get when they see one level of organization emerging from another, as simple rules give rise to complex behavior.

If you want to put in too many special properties, whether at the level of the individual human being or at higher levels of organization, you'd be going far beyond the capacity of any model. First, the model would become too difficult to handle mathematically, and second, once the model ran you'd find it very difficult to understand the results. There's always a trade-off between the advantages of stripping down the rules—so that you get caricatures of human beings, let's say, but you also get operations you can carry out mathematically—and the advantages of putting in something more complex, more sophisticated, more applicable to this planet and to the human race. Of course, as computers get better and better, the whole game will become more sophisticated, but there will still be such a trade-off.

An interesting question about the behavior of complex adaptive systems is, What is required to move from one level to another? In Tom Ray's little artificial world of digital organisms, there are significant jumps, and with more elaborate models we'll be able to see even more significant changes in level of organization.

The tendency of the researchers is to crowd over at the mathematical end of the spectrum, where the rules are simple and they get enormous pleasure out of seeing complexity emerge, but that work will be difficult to use for scientific or policy purposes, and rather easy to misuse. One has to invest some effort in the other parts of the spectrum as well.

Furthermore, one has to proceed with caution, in that much mischief has been done in the world by exaggerating the role of scientific metaphor in human affairs. The science of economics provides an example: people have tried to apply a stripped-down version of economics to human affairs, omitting a great many values, a great many things of importance. You get society in the service of economics, instead of economics in the service of society. The Nazi racial theories are, of course, a horrible example of misapplying metaphors from science. Nineteenth-century ideas of social Darwinism are another example. We have to be careful when we use these stripped-down models—and even when we use more complicated models—not to take them too seriously but rather to use them as prostheses for the imagination, as sources of inspiration, as acknowledged metaphors. In that way I think they can be valuable.

I've never been eager to sell a particular kind of activity to others just because I'm engaged in it myself. I never tried to sell elementary-particle physics to people as a career, and I wouldn't try to sell the study of complex adaptive systems to anybody either. I think what is exciting is human culture as a whole. People may want to be painters or poets or historians or scientists of various kinds—field biologists or archaeologists or plectics theorists or elementary-particle experimentalists or astronomers or whatever. It is noteworthy, though, that people who work on simplicity and complexity—on plectics—are often capable of carrying out practical activities in a great many different fields.

Nevertheless, people doing transdisciplinary work have a lot of problems finding suitable employment, especially in academic life. The reason isn't merely prejudice but also the fact that all the mechanisms for judging excellence are set up in the narrow traditional disciplines. Peer-reviewed journals, academic departments, Ph.D. exams, professional societies, and so on, are typically organized along disciplinary lines. Of course, there are always phonies who cower on the boundaries between fields, so people aren't altogether unjustified in being wary of transdisciplinary work. Clearly, we need effective mechanisms for judging it.

In discussing plectics with audiences, I encourage people to see one panorama rather than a lot of separate disciplines: the various meanings of simplicity and complexity; complex nonadaptive systems in the physical sciences; the modern interpretation of quantum mechanics; the simplicity of the fundamental laws of physics—that is, the unified theory of all the particles and their interactions plus the boundary condition at the beginning of the expansion of the universe; complex adaptive systems in the life sciences, in the behavioral and social sciences, and in practical human affairs; computer-based complex adaptive systems, some of which can serve as crude models for natural complex adaptive systems; and so forth.

Also, I have found it necessary to discuss the notion of reductionism. People scream epithets at one another over this issue of reduction. I take what I think is the only sensible position, which is that of course the basic laws of physics are fundamental in the sense that all the other laws are built on them, but that doesn't mean you can derive all the other laws from the laws of physics, because you have to add in all the special features of the world that come from history and that underlie the other sciences. Physics and chemistry stem from the fundamental laws, although even there, in the complicated branches of physics and chemistry, the formulation of the appropriate questions involves a great deal of special additional information about particular conditions that don't obtain everywhere in the universe. In the center of the sun, there is no solid-state physics. In the very early universe, when matter was still mostly a quark soup, there was not even nuclear physics. So even those subjects involve, in a sense, more than just fundamental laws.

All the rest of the sciences depend heavily on particular accidents in the history of the universe: astronomical accidents, geological accidents, biological accidents, accidents of human history, and so on. There's a huge body of information that has to be supplied in addition to the fundamental laws before you get the details of biology on Earth, for example. Just because elementary-particle physics is fundamental doesn't mean you can reduce biology to it, even in principle, unless you adjoin that additional information. Furthermore, in practice, it's essential to study biology at its own level, and likewise psychology, the social sciences, history, and so forth, because at each level you identify appropriate laws that apply at that level. Even though in principle those laws can be derived from the level below plus a lot of additional information, the reasonable strategy is to build staircases between levels both from the bottom up (with explanations in terms of mechanisms) and from the top down (with the discovery of important empirical laws). All of these ideas belong to what I call the doctrine of "emergence."

I've now retired from Caltech, an institution that is often labeled "reductionist," meaning that Caltech researchers usually don't take up in any depth subjects such as linguistics, archaeology, evolutionary biology, and psychology. Typically, they concentrate on fields like neurobiology, trying to investigate the mechanisms that underlie psychology. In that way, Caltech has built up a brilliant record of achievement in certain fields. However, in stressing the search for mechanism, Caltech tends to ignore the other part of strategy, which is to look for empirical rules in complicated fields and build staircases from the top down as well as from the bottom up.

Take Darwin, for example: would Caltech have hired Darwin? Probably not. He had only vague ideas about some of the mechanisms underlying biological evolution. He had no way of knowing about genetics, and he lived before the discovery of mutations. Nevertheless, he did work out, from the top down, the notion of natural selection and the magnificent idea of the relationship of all living things.

At the Santa Fe Institute, we encourage not only the study of plectics but also a number of general habits of research: building staircases from the top as well as the bottom, having the courage to take a crude look at the whole, cooperation among disciplines, and cooperation among different points of view on the same question when they are not logically contradictory. And we would have loved to have Darwin on our faculty.

CHRISTOPHER G. LANGTON: There's nothing like having a Nobel laureate around to liven up discussions on almost any topic. Often, however, receiving a Nobel Prize in one field gives the recipient the feeling that anything he or she says on any topic is worth listening to, which is generally not the case—with one howling exception: Murray.

Murray really *is* an expert in a wide variety of fields and really *does* know what he's talking about when he launches into a discourse on any one of them. He's probably fluent in as many scientific disciplines as he is in languages of the world, and I've lost count of how many languages he speaks. Sometimes it can be hard to get a conversation going off in a direction that doesn't include a topic that Murray's interested in, but the conversation will certainly never be dull or uninformative. I always learn a lot when I talk with Murray. I also have to say that Murray played a major role in setting up the intellectual atmosphere of the Santa Fe Institute, and he has been a strong advocate of the institute policy of reaching out to and including bright young researchers in addition to the more established older scientists who typically visit here.

ALAN GUTH: Murray Gell-Mann is certainly one of the three leading particle theorists of the century, along with Richard Feynman and Steven Weinberg. One of Murray's most important contributions was the discovery of the quark model. All particle physicists are now convinced that the so-called strongly interacting particles, which include the proton, neutron, and several hundred other particles less well known to the public, are all made out of fundamental constituents called quarks, and it was Murray who first proposed that. At the time, the evidence wasn't very strong; there were some patterns seen in the mass distribution of particles, but Murray put it all together and came up with the bold proposal that it would all look very simple if we assumed that these particles were made of quarks.

The proposal goes beyond that. It was not just a question of deciding that smaller particles existed—that by itself is kind of an obvious idea—but Murray went on to play an important role in constructing the detailed theory of how these quarks interact with each other, what their properties are, how you can use the properties of these quarks to calculate in detail the properties of the particles that the quarks make up. All that's very important; it's the backbone of our current understanding of particle physics, and Murray's role was absolutely crucial. The quark model became part of what's come to be called the standard model of particle physics, which is now the model that all of us accept.

The standard model is really a conglomeration of pieces that were developed by different people. The phrase "standard model" probably started to be used in 1974 or thereabouts. It's a phrase that caught on gradually, so it's a little hard to know when it was first used. The earliest piece of the standard model is the so-called electroweak theory, which was first published by Weinberg in 1967. The strong-interaction part of the standard model—the part about how quarks interact with one another—is based on papers that came out in 1971, 1972, and 1973, some of which were written by Gell-Mann.

We don't regard the standard model as the final theory; it's too complicated, too diverse in its description. Most particle theorists assume that the standard model is a low-energy approximation of a richer, fundamentally more simple theory. We have been looking for that more simple theory. Gell-Mann has played a role in that search, too; he wrote some of the important papers about grand unified theory—the unification of the electroweak and the strong interactions— when grand unified theories were first discussed. He also worked on some of the other ideas, like supergravity and superstrings.

Lately, Murray's gone off and done things I don't understand at all; he's left particle theory now, and he's working on complexity. Complexity remains a mystery to many particle theorists.

LEE SMOLIN: Murray is the greatest living American theoretical physicist. His contributions to elementary-particle theory were dramatic and very important. They came out of a tremendous imagination— the idea of strangeness, the idea of quarks, the idea of the eightfold way, the idea of SU(3).

SU(3) is the idea that all the known particles would be different manifestations of one kind of particle, and they'd be unified by a symmetry. A symmetry means a way of taking you from one particle to another particle—replacing one by another, in an experiment. The result of a symmetry is that the experiment is not much changed if you replace one particle by another. Murray's proposal was that there could be such a symmetry involving all the particles that were then known. This was in the early 1960s. The particles are of course not identical, but the idea is that the things that distinguish the particles would arise from smaller and less important effects than the things that made them similar, which could be explained with the notion of a symmetry. Symmetry is a profound idea that has been the driving force in elementary-particle physics since then. I'm not sure the idea is completely right, in the sense that it may have outlived its usefulness. But it's been the dominant idea since the 1960s.

Recently, Murray's been interested in more mathematical ideas. He played a big role in the establishment of the standard model; he was one of a number of people who pushed the idea that another sort of symmetry, called a gauge symmetry, could account for the forces that bind the quarks into the proton and neutron—this was quantum chromodynamics. He didn't invent supergravity, but he was important in its development. He invented a form of it with John Schwartz, and they played an important role in pushing the idea. Again, he didn't really make contributions to string theory, but he helped to push the idea. He also materially kept John Schwartz and some other string theorists alive and working as physicists for many years while nobody else was interested in strings. The fact that, after all this, he's become interested in the ideas of complexity is wonderful, because he's right: physics needs a new direction, and the direction should have something to do with the study of complex systems rather than with the kind of physics he did most of his life. The fact that after spending a life focused on studying the most elementary things in nature Murray can turn around and say that now what's important is the study of complex systems is a great inspiration, and also a great tribute to him.

What Murray is saying is that the important new ideas in science will come not from further development of particle physics in the direction of finding the perfect fundamental theory of everything, but in understanding why our universe is complex, and understanding how to mix the science of the fundamental with the science of the complex. It's a striking indication of his originality and intelligence that he's been thinking that way for a long time.

Murray also has ideas about the foundations of quantum mechanics and the interpretations of quantum mechanics and cosmology which are interesting, which have influenced a lot of people. I don't actually agree with these ideas—I have different ideas of my own—but certainly his ideas have played a big role in this field.

MARTIN REES: Great man. Clearly someone who has had remarkable success in predictions about particle physics over his career, and whose current work with the theoretical physicist Jim Hartle is influencing one of the main schools of thought in quantum gravity.

What Murray Gell-Mann appreciates is the contrast between the simplicity of particle physics and the complexity of the world around us. Quite different styles of thinking are needed for these kinds of phenomena. As a cosmologist, I like to describe the history of the universe in three parts. The first part is the first microsecond, which is difficult to understand because the basic microphysics is uncertain, involving extreme conditions that we can't replicate in accelerators. After the first microsecond, the universe becomes, in a sense, an easy place to understand; we can make calculations about primordial helium, deuterium, lithium, and so on, and about the spectrum of background radiation. But that simplicity ends after a few million years, when the first structures condense out from the universe. In the third part of its history, the universe becomes a complex place, and it remains a complex place thereafter, not because the basic physical laws are uncertain but because the manifestation of the laws in nonlinear structures are very complex.

Everything from meteorology to biology is essentially complex manifestations of simple laws. Most theoretical cosmologists are concerned with the early universe, where the laws are simple and there are no structures. That's a subject which is akin to particle physics, one side of Murray's interest. But the kind of cosmology I do (what some people call cosmogony, the study of the origin of the structures and of why the universe is the way it is) involves the emergence of complexity after the first few million years, after the fireball cools down. The nature of the subject then becomes different. We can't expect to encapsulate everything in a few simple equations, as in particle physics. We can't aspire to much beyond a qualitative understanding of some key processes. In that sense, it's more like the environmental sciences than like particle physics.

Murray Gell-Mann is someone who has emphasized this contrast but who appreciates the scientific challenges of both. That's one thing which is very admirable about him. Particle physicists have often been ultra-élitist, regarding their subject as being the highest paradigm, towards which all other sciences should strive. Murray is now emphasizing clearly that many other sciences are equally difficult and challenging, because of complexity. There is continued debate about whether some sciences are more fundamental and difficult than others, and it may be a mistake to regard the most fundamental sciences as being the most mathematical ones. Particle physics is actually a rather atypical science, in that it's the only science where you can expect things to be exactly described by a few equations. You don't expect continental drift to be described by a few equations; you expect a few unifying ideas.

In the particle-physics community, there are an enormous number of practitioners chasing a few key problems, and so if someone like Gell-Mann in the old days (or Ed Witten now) comes up with a key idea, lots of bright people follow its consequences very quickly. In astrophysics and cosmology, the ratio of bright people to problems is much lower. What that means is that often the good ideas not only don't get worked to death, they don't even get followed up enough. The frontiers are more extensive and less intensively developed, as it were.

J. DOYNE FARMER: The first thing that makes me respect Murray is that unlike all his contemporaries, including Feynman, Weinberg, Hawking, and all the other particle physicists, he saw that complexity is the next big problem. The kind of breakthroughs he made in the early 1960s in terms of impact on the world of science are not going to get made in that domain, they are going to get made in this domain. Murray recognized that, and has become more than just conversant with what's going on and with what the problems are.

What's impressed me is that when I heard Murray give his first few talks on complex systems, I thought he was missing the boat. Then I heard him speaking about it a few years later, and I thought he was accurately describing the boat. Murray is doing the field a great service by lending his name in support of it, and championing the cause, and he's also doing a good job of articulating what the cause is.

DANIEL C. DENNETT: Murray strikes me as having excellent instincts, scientifically. It's odd for me, as a philosopher, to praise scientists for having excellent scientific instincts, but I'm impressed with the fact that when he leaps into a controversy, his take on it is usually pretty apt. It always fascinates me to see how often fine scientists have a blinkered view of the world which prevents them from seeing the virtue of a certain approach. No blinkers on Murray.

STUART KAUFFMAN: Murray is enormously smart, sensible, and knowledgeable. He may know more things than any other single human being. He has played an extremely important role at the Santa Fe Institute in two or three guises. First of all, Murray's taste in science is good. His taste in people is good, too, even though he sometimes has a hard time expressing approval. He's been a continuous source of pressure toward broadening the institute and getting it to take on a wider range of issues. Secondly, Murray has lent enormous prestige to whatever the sciences of complexity will be. He's laid his reputation on the line in helping to found the institute and being out there as a public spokesman for what we're doing. Thirdly, while Murray has obviously dominated physics for years, in the emerging sciences of complexity he hasn't made major contributions of an original kind. What he has done is to assemble what are essentially other people's ideas into his own coherent framework.

MARVIN MINSKY: What is there to say? He's wonderful. He's right up there with Feynman as one of the great thinkers. He knows a lot about many things, including artificial intelligence. But I think his major contribution is inventing new kinds of insults. For instance, if somebody says something that isn't exactly perfect—Murray has developed one of the best inventories of put-downs that exists. I hear he's getting mellower. That would be a terrible loss for civilization. A collection of anecdotes about his remarks about other people would be priceless.

PAUL DAVIES: Murray Gell-Mann is one of the towering figures in twentieth-century physics. He'll go down in history as the founder, or one of the cofounders, of the idea of quarks, the elementary constituents of the nuclear particles. It's only in recent years that he's become known for his work on complexity theory. What he's done is to recognize the fact that there are two ways of studying the world. There's the reductionist path, in which you try to break things down into their most elementary constituents—quarks, or maybe something deeper, like superstrings. The other path is the path of synthesis, the path of looking at the complex organizational arrangement of things and recognizing that there's a whole science of complexity, with laws and principles emerging at successive levels.

Chapter 20

STUART KAUFFMAN

"Order for Free"

BRIAN GOODWIN: Stuart is primarily interested in the emergence of order in evolutionary systems. That's his fix. It's exactly the same as mine, in terms of the orientation towards biology, but he uses a very different approach. Our approaches are complementary with respect to the same problem: How do you understand emergent novelty in evolution? Emergent order? Stuart's great contributions are there.

•   •   •

STUART KAUFFMAN is a biologist; professor of biochemistry at the University of Pennsylvania and a professor at the Santa Fe Institute; author of Origins of Order: Self-Organization and Selection in Evolution *(1993), and co-author with George Johnson of* At Home in the Universe *(1995).*

STUART KAUFFMAN: What kinds of complex systems can evolve by accumulation of successive useful variations? Does selection by itself achieve complex systems able to adapt? Are there lawful properties characterizing such complex systems? The overall answer may be that complex systems constructed so that they're on the boundary between order and chaos are those best able to adapt by mutation and selection.

Chaos is a subset of complexity. It's an analysis of the behavior of continuous dynamical systems—like hydrodynamic systems, or the weather—or discrete systems that show recurrences of features and high sensitivity to initial conditions, such that very small changes in the initial conditions can lead a system to behave in very different ways. A good example of this is the so-called butterfly effect: the idea is that a butterfly in Rio can change the weather in Chicago. An infinitesimal change in initial conditions leads to divergent pathways in the evolution of the system. Those pathways are called trajectories. The enormous puzzle is the following: in order for life to have evolved, it can't possibly be the case that trajectories are always diverging. Biological systems can't work if divergence is all that's going on. You have to ask what kinds of complex systems can accumulate useful variation.

We've discovered the fact that in the evolution of life very complex systems can have convergent flow and not divergent flow. Divergent flow is sensitivity to initial conditions. Convergent flow means that even different starting places that are far apart come closer together. That's the fundamental principle of homeostasis, or stability to perturbation, and it's a natural feature of many complex systems.

We haven't known that until now. That's what I found out twenty-five years ago, looking at what are now called Kauffman models—random networks exhibiting what I call "order for free."

Complex systems have evolved which may have learned to balance divergence and convergence, so that they're poised between chaos and order. Chris Langton has made this point, too. It's precisely those systems that can simultaneously perform the most complex tasks and evolve, in the sense that they can accumulate successive useful variations. The very ability to adapt is itself, I believe, the consequence of evolution. You have to be a certain kind of complex system to adapt, and you have to be a certain kind of complex system to coevolve with other complex systems. We have to understand what it means for complex systems to come to know one another—in the sense that when complex systems coevolve, each sets the conditions of success for the others. I suspect that there are emergent laws about how such complex systems work, so that, in a global, Gaia-like way, complex coevolving systems mutually get themselves to the edge of chaos, where they're poised in a balanced state. It's a very pretty idea. It may be right, too.

My approach to the coevolution of complex systems is my order-for-free theory. If you have a hundred thousand genes and you know that genes turn one another on and off, then there's some kind of circuitry among the hundred thousand genes. Each gene has regulatory inputs from other genes that turn it on and off. This was the puzzle: What kind of a system could have a hundred thousand genes turning one another on and off, yet evolve by creating new genes, new logic, and new connections?

Suppose we don't know much about such circuitry. Suppose all we know are such things as the number of genes, the number of genes that regulate each gene, the connectivity of the system, and something about the kind of rules by which genes turn one another on and off. My question was the following: Can you get something good and biology-like to happen even in randomly built networks with some sort of statistical connectivity properties? It can't be the case that it has to be very precise in order to work—I hoped, I bet, I intuited, I believed, on no good grounds whatsoever—but the research program tried to figure out if that might be true. The impulse was to find order for free. As it happens, I found it. And it's profound.

One reason it's profound is that if the dynamical systems that underlie life were inherently chaotic, then for cells and organisms to work at all there'd have to be an extraordinary amount of selection to get things to behave with reliability and regularity. It's not clear that natural selection could ever have gotten started without some preexisting order. You have to have a certain amount of order to select for improved variants.

Think of a wiring diagram that has ten thousand light bulbs, each of which has inputs from two other light bulbs. That's all I'm going to tell you. You pick the inputs to each bulb at random, and put connecting wires between them, and then assign one of the possible switching rules to each of the light bulbs at random. One rule might be that a light bulb turns on at the next moment if both of its inputs are on at the previous moment. Or it might turn on if both of its inputs are off.

If you go with your intuition, or if you ask outstanding physicists, you'll reach the conclusion that such a system will behave chaotically. You're dealing with a random wiring diagram, with random logic—a massively complex, disordered, parallel-processing network. You'd think that in order to get such a system to do something orderly you'd have to build it in a precise way. That intuition is fundamentally wrong. The fact that it's wrong is what I call "order for free."

There are other epistemological considerations regarding "order for free." In the next few years, I plan to ask, "What do complex systems have to be so that they can know their worlds?" By "know" I don't mean to imply consciousness; but a complex system like the *E. coli* bacterium clearly knows its world. It exchanges molecular variables with its world, and swims upstream in a glucose gradient. In some sense, it has an internal representation of that world. It's also true that IBM in some sense knows its world. I have a hunch that there's some deep way in which IBM and *E. coli* know their worlds in the same way. I suspect that there's no one person at IBM who knows IBM's world, but the organization gets a grip on its economic environment. What's the logic of the structure of these systems and the worlds that they come to mutually live in, so that entities that are complex and ordered in this way can successfully cope with one another? There must be some deep principles.

For example, IBM is an organization that knows itself, but I'm not quite talking about Darwinian natural selection operating as an outside force. Although Darwin presented natural selection as an external force, what we're thinking of is organisms living in an environment that consists mostly of other organisms. That means that for the past four billion years, evolution has brought forth organisms that successfully coevolved with one another. Undoubtedly natural selection is part of the motor, but it's also true that there is spontaneous order.

By spontaneous order, or order for free, I mean this penchant that complex systems have for exhibiting convergent rather than divergent flow, so that they show an inherent homeostasis, and then, too, the possibility that natural selection can mold the structure of systems so that they're poised between these two flows, poised between order and chaos. It's precisely systems of this kind that will provide us with a macroscopic law that defines ecosystems, and I suspect it may define economic systems as well.

While it may sound as if "order for free" is a serious challenge to Darwinian evolution, it's not so much that I want to challenge Darwinism and say that Darwin was wrong. I don't think he was wrong at all. I have no doubt that natural selection is an overriding, brilliant idea and a major force in evolution, but there are parts of it that Darwin couldn't have gotten right. One is that if there is order for free—if you have complex systems with powerfully ordered properties—you have to ask a question that evolutionary theories have never asked: Granting that selection is operating all the time, how do we build a theory that combines self-organization of complex systems—that is, this order for free—and natural selection? There's no body of theory in science that does this. There's nothing in physics that does this, because there's no natural selection in physics—there's self-organization. Biology hasn't done it, because although we have a theory of selection, we've never married it to ideas of self-organization. One thing we have to do is broaden evolutionary theory to describe what happens when selection acts on systems that already have robust self-organizing properties. This body of theory simply does not exist.

There are a couple of parallels concerning order for free. We've believed since Darwin that the only source of order in organisms is selection. This is inherent in the French biologist François Jacob's phrase that organisms are "tinkered-together contraptions." The idea is that evolution is an opportunist that tinkers together these widgets that work, and the order you see in an organism has, as its source, essentially only selection, which manages to craft something that will work. But if there's order for free, then some of the order you see in organisms is not due to selection. It's due to something somehow inherent in the building blocks. If that's right, it's a profound shift, in a variety of ways.

The origin of life might be another example of order for free. If you have complex-enough systems of polymers capable of catalytic action, they'll self-organize into an autocatalytic system and, essentially, simply *be* alive. Life may not be as hard to come by as we think it is.

There are some immediate possibilities for the practical application of these theories, particularly in the area of applied molecular evolution. In 1985, Marc Ballivet and I applied for a patent based on the idea of generating very, very large numbers of partly or completely random DNA sequences, and therefrom RNA sequences, and from that proteins, to learn how to evolve biopolymers for use as drugs, vaccines, enzymes, and so forth. By "very large" I mean numbers on the order of billions, maybe trillions of genes—new genes, ones that have never before existed in biology. Build random genes, or partly random genes. Put them into an organism. Make partly random RNA molecules; from that make partly random proteins, and learn from that how to make drugs or vaccines. Within five years, I hope we'll be able to make vaccines to treat almost any disease you want, and do it rapidly. We're going to be able to make hundreds of new drugs.

A related area is that probably a hundred million molecules would suffice as a roughed-in universal toolbox, to catalyze any possible reaction. If you want to catalyze a specific reaction, you go to the toolbox, you pull out a roughed-in enzyme, you tune it up by some mutations, and you catalyze any reaction you want. This will transform biotechnology. It will transform chemistry.

There are also connections to be made between evolutionary theory and economics. One of the fundamental problems in economics is that of bounded rationality. The question in bounded rationality is, How can agents who aren't infinitely rational and don't have infinite computational resources get along in their worlds? There's an optimizing principle about precisely how intelligent such agents ought to be. If they're either too intelligent or too stupid, the system doesn't evolve well.

Economist colleagues and I are discussing the evolution of a technological web, in which new goods and services come into existence and in which one can see bounded rationality in a nonequilibrium theory of price formation. It's the next step toward understanding what it means for complex systems to have maps of their world and to undertake actions for their own benefit which are optimally complex or optimally intelligent—boundedly rational. It's also part of the attempt to understand how complex systems come to know their world.

BRIAN GOODWIN: Stuart is primarily interested in the emergence of order in evolutionary systems. That's his fix. It's exactly the same as mine, in terms of the orientation toward biology, but he uses a very different approach. Our approaches are complementary with respect to the same problem: How do you understand emergent novelty in evolution? Emergent order? Stuart's great contributions are there.

The notion of life at the edge of chaos is absolutely germane to Stuart's work. He didn't discover that phrase, but his work has always been concerned with precisely that notion, of how you have an immensely complex system with patterns of interaction that don't obviously lead anywhere, and suddenly out pops order.

That's what he discovered when he was a medical student in the sixties messing about with computers. He worked with François Jacob's and Jacques Monod's ideas about controls. He implemented those on his computer, and he looked at neural networks. It's the same thing that inspired me, but we went in different directions. I went in the direction of the organism as a dynamic organization, and he was much closer to Warren McCulloch and the notion of logical networks and applying it to gene networks. Stuart and I have always had this complementary approach to things, and yet we come to exactly the same conclusions about the emergence of order out of chaotic dynamics. Stuart has the fastest flow of interesting new ideas of anybody I've ever met. I've learned a lot from him.

W. DANIEL HILLIS: Stuart Kauffman is a strange creature, because he's a theoretical biologist, which is almost an oxymoron. In physics, there are the theoretical types and the experimental types, and there's a good understanding of what the relationship is between them. There's a tremendous respect for the theoreticians. In physics, the theory is almost the real stuff, and the experiments are just an approximation to test the theory. If you get something a little bit wrong, then it's probably an experimental error. The theory is the thing of perfection, unless you find an experiment that shows that you need to shift to another theory. When Eddington went off during a solar eclipse to measure the bending of starlight by the sun and thus to test Einstein's general-relativity theory, somebody asked Einstein what he would think if Eddington's measurements failed to support his theory, and Einstein's comment was, "Then I would have felt sorry for the dear Lord. The theory is correct."

In biology, however, this is reversed. The experimental is on top, and the theory is considered poor stuff. Everything in biology is data. The way to acquire respect is to spend hours in the lab, and have your students and postdocs spend hours in the lab, getting data. In some sense, you're not licensed to theorize unless you get the data. And you're allowed to theorize only about your own data—or at the very least you need to have collected data before you get the right to theorize about other data.

Stuart is of the rare breed that generates theories without being an experimentalist. He takes the trouble to understand things, such as dynamical-systems theory, and tries to connect those into biology, so he becomes a conduit of ideas that are coming out of physics, from the theorists in physics, into biology.

DANIEL C. DENNETT: Stuart Kauffman and his colleague Brian Goodwin are particularly eager to discredit the powerful image first made popular by the great French biologists Jacques Monod and François Jacob—the image of Mother Nature as a tinkerer engaged in the opportunistic handiwork that the French call *bricolage*. Kauffman wants to stress that the biological world is much more a world of Newtonian discoveries than of Shakespearean creations. He's certainly found some excellent demonstrations to back up this claim. Kauffman is a meta-engineer. I fear that his attack on the metaphor of the tinkerer feeds the yearning of those who don't appreciate Darwin's dangerous idea. It gives them a false hope that they're seeing not the forced hand of the tinkerer but the divine hand of God in the workings of nature. Kauffman gets that from Brian Goodwin. John Maynard Smith has been pulling Kauffman in the other direction—very wisely so, in my opinion.

STEPHEN JAY GOULD: Stuart Kauffman is very similar to Brian Goodwin, in that they are both trying to explore the relevance of the grand structuralist tradition, which Darwinian functionalism never paid a whole lot of attention to. Stuart is different from Brian, in that Brian focuses upon the morphology of organisms. Stuart's main interests are in questions of the origin of life, the origins of molecular organization, which I don't understand very well. I'm not as quantitative as he is, so I don't follow all the arguments in his book. He's trying to understand what aspects of organic order follow from the physical principles of matter, and the mathematical structure of nature, and need not be seen as Darwinian optimalities produced by natural selection.

He's following in the structuralist tradition, which should not be seen as contrary to Darwin but as helpful to Darwin. Structural principles set constraints, and natural selection must work within them. His "order for free" is an outcome of sets of constraints; it shows that a great deal of order can be produced just from the physical attributes of matter and the structural principles of organization. You don't need a special Darwinian argument; that's what he means by "order for free." It's a very good phrase, because a strict Darwinian thinks that all sensible order has to come from natural selection. That's not true.

J. DOYNE FARMER: Stuart Kauffman was in a theoretical-biology group at the University of Chicago, run by Jack Cowan, that included people like Arthur Winfree, Leon Glass, and several others who have become some of the most famous theoretical biologists. The fact that any of these guys are still employed as scientists is a tribute to their ability; most of the biology establishment hates theoreticians and surviving as a theoretical biologist is difficult. Stuart survived, in part, by doing experiments as well, but I think his real passion has always been for theoretical biology.

FRANCISCO VARELA: Stuart has taken the notion of seeing emerging levels in biological organizations into explicit forms and mechanisms. In his early work on genetic networks, he did some very fundamental things. He took something that was vague and made it into a concrete example that was workable.

I have a little harder time with his last book. The monster, *The*

Origins of Order. Although many of the pieces in there have a flavor of something quite interesting, it doesn't seem to me that the book hangs together as a whole. There's too much of "Let's assume this, and let's assume that, and if this were right, then. . . ." But the basic idea is that we're back to the notion of evolution having intrinsic factors, and in this regard it has to be right. It's like Nick Humphrey's book. Although the smell is the right one, I'm not so sure I can buy the actual theory that he's trying to stitch together.

Stuart is one of the most competent people we have around when it comes to dealing with molecular biological networks. He's one of the great people, in that he has put some important bricks in that edifice, but that edifice has been built by many other people as well: Gould, Eldredge, Margulis, Goodwin. If there's a slight criticism I would make of Stuart, it's that sometimes he's not so clear in acknowledging that. What's happening here is that there's an evolution—or revolution—in biology, which is going beyond Darwin. But this revolution is not reducible to Stuart's own way of expressing it.

NILES ELDREDGE: Stuart is amazing. He had me on the floor of a cab, doubled up in laughter, the first time I met him. He was imitating all of the variant accents of the Oxford dons in philosophy. He's an amazingly funny guy, very likable guy, and extremely bright, of course. He takes what I used to call a transformationalist approach to evolution.

The standard way of looking at evolution is that evolution is a matter of transforming the physical properties of organisms. Stuart's got models jumping around from adaptive peak to adaptive peak, to explain the early Cambrian explosion. There's so much missing between the way he's thinking about things and the way I'm thinking about things that we've never really connected. We've talked, and I've put him together with other people who use computers to simulate evolutionary patterns, but there's just too much of a gap in our approach to things for there to be much useful dialog between us.

NICHOLAS HUMPHREY: Kauffman is less radical than Goodwin, at least nowadays. Kauffman originally would have said that natural selection doesn't play a very important role, but he's been persuaded that even if the possibilities that biology has to play with are determined by the properties of complex systems, nonetheless the ones we see in na-

STUART KAUFFMAN *"Order for Free"*   343

ture are those that have been selected. The world throws up possibilities, and then natural selection gets to work and ensures that just certain ones survive.

Kauffman is doing wonderful work, and he's certainly put the cat among the pigeons for old-fashioned neo-Darwinism. He's forced people to recognize that selection may not be the only designing force in nature. But he's not claiming to be the new Darwin. We don't need a new Darwin.

Chapter 21

CHRISTOPHER G. LANGTON

"A Dynamical Pattern"

W. DANIEL HILLIS: *Chris Langton is the central guru of this artificial-life stuff. He's onto a good idea when he says that life seems to be at the transition between order and disorder, as he calls it: right at the edge of chaos, just at the temperature between where water is ice and where water is steam, that area where it's liquid—right in between. In many ways, we're poised on the edge between being too structured and too unstructured.*

• • •

CHRISTOPHER G. LANGTON is a computer scientist; visiting professor at the Santa Fe Institute; director of the institute's artificial-life program; editor of the journal Artificial Life; author of Artificial Life (1995).

CHRISTOPHER G. LANGTON: What was Darwin's algorithm? The idea of evolution had been around for a long time. Spencer, Lamarck, and others had proposed evolution as a process, but they didn't have a mechanism. The problem was revealing the mechanism—the algorithm—that would account for the tremendous diversity observed in nature, in all its scope and detail. The essence of an algorithm is the notion of a finitely specified, step-by-step procedure to resolve a set of inputs into a set of outputs. Darwin's genius was to take the huge variety of species he saw on the planet and propose a simple, elegant mechanism, a step-by-step procedure, that could explain their existence.

Darwin distinguished two fundamental roles: there had to be (1) a producer of variety and (2) a filter of variety. In the first few chapters of *Origin of Species*, Darwin appeals to his contemporaries' common knowledge that nature produces variability in the offspring of organisms. Everybody knew about the breeding of plants and animals. It was clear that the variety depended upon by breeders was a product of natural processes going on in animal and plant reproduction. A human breeder could arrange specific matings to take advantage of this natural variability to enhance certain desired traits among his stock. One could say, "Those two sheep produce more wool than most of the others, so I'm going to mate those two and get sheep that produce more wool." Although the variety was produced naturally, a human breeder arranged for the matings. Since the "filter" of the variety was an artifact of human design, this process is termed artificial selection.

Having cast the situation in terms familiar to his contemporaries,

Darwin devoted the remainder of his book to showing how "Nature Herself" could fulfill the role of the selective filter: the entity that arranged for certain matings to take place preferentially, based on the traits of the individuals involved. Since certain traits enhanced the likelihood of survival for the organisms that bore them, organisms carrying those traits would be more likely to survive and mate than organisms that didn't carry those traits.

It's possible to cast this process in terms of a step-by-step procedure called a genetic algorithm, which runs on a computer, allowing us to abstract the process of evolution from its material substrate. John Holland, of the University of Michigan, was the first to seriously pursue implementing Darwin's algorithm in computers in the early 1960s.

People have been working with genetic algorithms ever since, but these algorithms haven't been very useful tools for studying biological evolution. This isn't because there's anything wrong with the algorithms per se, but rather because they haven't been embedded in the proper biological context. As genetic algorithms have been traditionally implemented, they clearly involve artificial selection: some human being provides explicit, algorithmic criteria for which of the entities is to survive to mate and reproduce. The real world, however, makes use of natural selection, in which it is the "nature" of the interactions among all the organisms—both with one another and with the physical environment—that determines which entities will survive to mate and reproduce. It required a bit of experimentation to work out how to bring about natural selection within the artificial worlds we create in computers.

Over the last several years, however, we've learned how to do that, through the work of Danny Hillis, Tom Ray, and others. We don't specify the selective criteria externally. Rather, we let all the "organisms" interact with one another, in the context of a dynamic environment, and the selective criteria simply emerge naturally. To any one of these organisms, "nature," in the computer, is the collective dynamics of the rest of the computerized organisms there. When we allow this kind of interaction among the organisms—when we allow them to pose their own problems to one another—we see the emergence of a Nature with a capital "N" inside the computer, whose "nature" we can't predict as it evolves through time.

Typically, a collection of organisms in such artificial worlds will form an ecology, which will be stable for a while but will ultimately collapse. After a chaotic transition, another stable ecology will form, and the process continues. What defines fitness—and what applies the selective pressure—is this constantly changing collective activity of the set of organisms themselves. I argue that such a virtual ecosystem—what I have termed "artificial life"—constitutes a genuine "nature under glass," and that the study of these virtual natures within computers can be extremely useful for studying the nature of nature outside the computer.

The notion of a human-created nature in a computer can be a little perplexing to people at first. Computers run algorithms, and algorithms seem to be in direct contrast to the natural world. The natural world tends to be wild, woolly, and unpredictable, while algorithms tend to be precise, predictable, and understandable. You know the outcome of an algorithm; you know what it's going to do, because you've programmed it to do just that. Because algorithms run on computers, you expect the "nature" of what goes on in computers to be as precise and predictable as algorithms appear to be. However, those of us who have a lot of experience with computers realize that even the simplest algorithms can produce completely novel and totally unpredictable behaviors. The world inside a computer can be every bit as wild and woolly as the world outside.

One can think of a computer in two ways: as something that runs a program and calculates a number, or as a kind of logical digital universe that behaves in many different ways. At the first artificial-life workshop, which I organized at Los Alamos National Laboratory in 1987, we asked ourselves, How are people going about modeling living things? How are we going about modeling evolution, and what problems do we run into? Once we saw the ways everyone was approaching these problems, we realized that there was a fundamental architecture underlying the most interesting models: they consisted of many simple things interacting with one another to do something collectively complex. By experimenting with this distributed kind of computational architecture, we created in our computers universes that were complex enough to support processes that, with respect to those universes, have to be considered to be alive. These processes behave in their universes the way living things behave in our universe.

I don't see artificial intelligence and artificial life as two distinct enterprises in principle; however, they're quite different in practice. Both endeavors involve attempts to synthesize—in computers—natural processes that depend vitally on information processing. I find it hard to draw a dividing line between life and intelligence. Both AI and AL study systems that determine their own behavior in the context of the information processes inside them. AI researchers picked the most complex example in that set, human beings, and were initially encouraged—and misled—by the fact that it appeared to be easy to get computers to do things that human beings consider hard, like playing chess. They met with a lot of initial success at what turned out to be not very difficult problems. The problems that turned out to be hard were, ironically enough, those things that seem easy to human beings, like picking out a friend's face in a crowd, walking, or catching a baseball. By contrast, artificial-life researchers have decided to focus on the simplest examples of natural information processors, such as single cells, insects, and collections of simple organisms like ant colonies.

Our approach to the study of life and, ultimately, intelligence and consciousness is very bottom-up. Rather than trying to describe a phenomenon at its own level, we want to go down several levels to the mechanisms giving rise to it, and try to understand how the phenomenon emerges from those lower-level dynamics. For instance, fluid dynamics is reasonably well described explicitly by Navier-Stokes equations, but this is a high-level description imposed on the system from the outside and from the top down; the fluid itself does not compute Navier-Stokes equations. Rather, the fluid's behavior emerges out of interactions between all of the particles that make it up—for example, water molecules. Thus, one can also capture fluid dynamics implicitly, in the rules for collisions among the particles of which a fluid is constituted. The latter approach is the bottom-up approach, and it's truer to the way in which behavior is generated in nature. The traditional AI approach to intelligence is akin to the Navier-Stokes approach to fluid dynamics. However, in the case of phenomena like life and intelligence, we haven't been able to come up with high-level, top-down "rules" that work. In my opinion, this is more than just a case of not yet having found them; I think it's quite likely that no such rules can be formulated.

In the early days of artificial intelligence, researchers assumed that the most important thing about the brain, for the purposes of understanding intelligence, was that it was a universal computer. Its parallel, distributed architecture was thought to be merely a consequence of the bizarre path that nature had to take to evolve a universal computer. Since we know that all universal computers are equivalent in principle in their computational power, it was thought that we could effectively ignore the architecture of the brain and get intelligent software running on our newly engineered universal computers, which had very different architectures. However, I think that the difference in architecture is crucial. Our engineered computers involve a central controller working from a top-down set of rules, while the brain has no such central controller and works in a very distributed, parallel manner, from the bottom up. What's natural and spontaneous for this latter architecture can be achieved by the former only by using our standard serial computers to simulate parallel, distributed systems. There's something in the dynamics of parallel, distributed, highly nonlinear systems which lies at the roots of intelligence and consciousness—something that nature was able to discover and take advantage of.

What trick is it that nature capitalized on in order to create consciousness? We don't yet understand it, and the reason is that we don't understand what very distributed, massively parallel networks of simple interacting agents are capable of doing. We don't have a good feel for what the spectrum of possible behaviors is. We need to chart them, and once we do we may very well discover that there are some phenomena we didn't know about before—phenomena that turn out to be critical to understanding intelligence. We won't discover them if we work from the top down.

If you look at the architecture of most of the complex systems in nature—immune systems, economies, countries, corporations, living cells—there's no central controller in complete control of the system. There may be things that play a slightly centralized role, such as the nucleus in a cell, or a central government, but a great deal of the dynamics goes on autonomously. In fact, many of the emergent properties that such systems get caught up in would probably not be possible if everything had to be controlled by a centralized set of rules. Nature has learned how to bring about organization without employing a central organizer, and the resulting organizations seem much more robust, adaptive, flexible, and innovative than those we build ourselves that rely on a central controller.

In fact, natural systems didn't evolve under conditions that particularly favored central control. Anything that existed in nature had to behave in the context of a zillion other things out there behaving and interacting with it, without any one of these processes gaining control of the whole system and dictating to the others what to do. This is a very distributed, massively parallel architecture.

Think of an ant colony—a beautiful example of a massively parallel, distributed system. There's no one ant that's calling the shots, picking from among all the other ants which one is going to get to do its thing. Rather, each ant has a very restricted set of behaviors, but all the ants are executing their behaviors all the time, mediated by the behaviors of the ants they interact with and the state of the local environment. When one takes these behaviors on aggregate, the whole collection of ants exhibits a behavior, at the level of the colony itself, which is close to being intelligent. But it's not because there's an intelligent individual telling all the others what to do. A collective pattern, a dynamical pattern, takes over the population, endowing the whole with modes of behavior far beyond the simple sum of the behaviors of its constituent individuals. This is almost vitalistic, but not quite, because the collective pattern has its roots firmly in the behavior of the individual ants.

This example shows how one can be both a vitalist and a mechanist at the same time. We have a set of interacting agents, and they run into one another and do things based on their local interactions. That microcosm gives rise to a collective pattern of global dynamics. In turn, these global patterns set the context within which the agents interact—a context that can be a fairly stabilizing force. If it's too stabilizing, however, the system freezes, like a crystal, and can no longer react in a dynamic way to external pressures. The system as a whole has to respond to external pressures more like a fluid than a crystal, and thus it must be the case that the patterns that emerge can be easily destabilized under appropriate conditions, to be replaced by patterns that are more stable under the new circumstances. It could be that even without external perturbations one pattern of activity will reign for a while and ultimately collapse, to be replaced by another pattern—a stable organization under the new conditions. So, global patterns of organization can be causal, just as the vitalist wants, but these very patterns depend on the dynamics of the microcosm they inform, and don't exist independently of the entities that make up that microcosm, just as the mechanist requires.

In the late nineteenth century, the Austrian physicist Ludwig Boltzmann showed that one could account for many of the thermodynamic properties of macroscopic systems in terms of the collective activity of their constituent atoms. Boltzmann's most famous contribution to our understanding of the relationship between the microcosm of atoms and the macroscopic world of our experience was his definition of entropy: $S = k \, log \, W$. In the 1950s, the computer scientist Claude Shannon generalized Boltzmann's formula, lifting the concept of entropy from the thermodynamic setting in which it was discovered to the more general level of probability theory, providing a precise, quantitative meaning for the term "information." That was a good start. But a lot more needs to be lifted from the domain of thermodynamics. Other useful quantities to generalize from thermodynamics include energy and temperature. I'm convinced that generalizing other concepts from thermodynamics and statistical mechanics will have a major impact on our understanding of biology and other complex systems.

As Doyne Farmer has pointed out, our current understanding of complex systems is very much in the same state as our understanding of thermodynamics was in the mid-1800s, when people were screwing around with the basic concepts but didn't yet know which were the right quantities to measure. Until you know which are the relevant quantities to measure, you can't come up with quantitative expressions relating those quantities to one another. The French physicist Sadi Carnot was one of the first people to identify some basic quantities, such as heat and work. He was followed by a stream of people, like Rudolf Clausius and Josiah Willard Gibbs, until Boltzmann finally made the connection between the microcosmos of atoms and the macrocosmos of thermodynamics.

In my own work, I've focused on some general properties of thermodynamic systems which appear to be important for understanding complex systems. There are certain regimes of behavior of physical systems, generally called "phase transitions," which are best characterized by statistical mechanics. A physical system undergoes a phase transition when its state changes—for instance, when water freezes into ice. I've found that during phase transitions physical systems often exhibit their most complex behavior. I've also found that it's during phase transitions that information processes can appear spontaneously in physical systems and play an important role in the determination of the systems' behavior. One might even say that systems at phase transitions are caught up in complex computations to determine their own physical state. My belief is that the dynamics of phase transitions are the point at which information processing can gain a foothold in physical systems, gaining the upper hand over energy in the determination of the systems' behavior. It has long been a goal of science to discover where and how information theory and physics fit into each other; it's become something of a Holy Grail. I can't say I've found the Grail, but I do think I've found the mountain range it's located in.

People have been trying to synthesize life for a long time, but in most cases they were trying to build models that were explicitly like the life we know. When people would build a model of life, it would be a model of a duck, or a model of a mouse. The Hungarian mathematician John Von Neumann had the insight that we could learn a lot even if we didn't try to model some specific existing biological thing. He went after the logical basis, rather than the material basis, of a biological process, by attempting to abstract the logic of self-reproduction without trying to capture the mechanics of self-reproduction (which were not known in the late 1940s, when he started his investigations).

Von Neumann demonstrated that one could have a machine, in the sense of an algorithm, that would reproduce itself. Most biologists weren't interested, because it wasn't like any specific instance of biological self-reproduction (it wasn't a model of chromosomes, for example). Von Neumann was able to derive some general principles for the process of self-reproduction. For instance, he determined that the information in a genetic description, whatever it was, had to be used in two different ways: (1) it had to be interpreted as instructions for constructing itself or its offspring, and (2) it had to be copied passively, without being interpreted. This turned out to be the case for the information stored in DNA when James Watson and Francis Crick determined its structure in 1953. It was a far-reaching and very prescient thing to realize that one could learn something about "real biology" by studying something that was not real biology—by trying to get at the underlying "bio-logic" of life.

That approach is characteristic of artificial life. AL attempts to look beyond the collection of naturally occurring life in order to discover things about that set that could not be discovered by studying that set alone. AL isn't the same thing as computational biology, which primarily restricts itself to computational problems arising in the attempt to analyze biological data, such as algorithms for matching protein sequences to gene sequences, or programs to reconstruct phylogenies from comparisons of gene sequences. Artificial life reaches far beyond computational biology. For example, AL investigates evolution by studying evolving populations of computer programs—entities that aren't even attempting to be anything like "natural" organisms.

Many biologists wouldn't agree with that, saying that we're only simulating evolution. But what's the difference between the process of evolution in a computer and the process of evolution outside the computer? The entities that are being evolved are made of different stuff, but the process is identical. I'm convinced that such biologists will eventually come around to our point of view, because these abstract computer processes make it possible to pose and answer questions about evolution that are not answerable if all one has to work with is the fossil record and fruit flies.

The idea of artificially created life is pregnant with issues for every branch of philosophy, be it ontology, epistemology, or moral or social philosophy. Whether it happens in the next ten, hundred, or only in the next thousand years, we are at the stage where it's become possible to create living things that are connected to us not so much by material as by information. In geological time, even a thousand years is an instant, so we're literally at the end of one era of evolution and at the beginning of another. It's easy to descend into fantasy at this point, because we don't know what the possible outcome of producing "genuine" artificial life will be. If we create robots that can survive on their own, can refine their own materials to construct offspring, and can do so in such a way as to produce variants that give rise to evolutionary lineages, we'll have no way of predicting their future or the interactions between their descendants and our own. There are quite a few issues we need to think about and address before we initiate such a process. A reporter once asked me how I would feel about my children living in an era in which there was a lot of artificial life. I answered, "Which children are you referring to? My biological children, or the artifactual children of my mind?"—to use Hans Moravec's phrase. They would both be my children, in a sense.

It's going to be hard for people to accept the idea that machines can be as alive as people, and that there's nothing special about our life that's not achievable by any other kind of stuff out there, if that stuff is put together in the right way. It's going to be as hard for people to accept that as it was for Galileo's contemporaries to accept the fact that Earth was not at the center of the universe. Vitalism is a philosophical perspective that assumes that life cannot be reduced to the mere operation of a machine, but, as the British philosopher and scientist C.H. Waddington has pointed out, this assumes that we know what a machine is and what it's capable of doing.

Another set of philosophical issues raised in the pursuit of artificial life centers on questions of the nature of our own existence, of our own reality and the reality of the universe we live in. After working for a long time creating these artificial universes, wondering about getting life going in them, and wondering if such life would ever wonder about its own existence and origins, I find myself looking over my shoulder and wondering if there isn't another level on top of ours, with something wondering about me in the same way. It's a spooky feeling to be caught in the middle of such an ontological recursion. This is Edward Fredkin's view: the universe as we know it is an artifact in a computer in a more "real" universe. This is a very nice notion, if only for the perspective to be gained from it as a thought experiment—as a way to enhance one's objectivity with respect to the reality one's embedded in.

Biology has until now been occupied with taking apart what's already alive and trying to understand, based on that, what life is. But we're finding that we can learn a lot by trying to put life together from scratch, by trying to create our own life, and finding out what problems we run into. Things aren't necessarily as simple—or, perhaps, as complicated—as we thought. Furthermore, the simple change in perspective—from the analysis of "what is" to the synthesis of "what could be"—forces us to think about the universe not as a given but as a much more open set of possibilities. Physics has largely been the science of necessity, uncovering the fundamental laws of nature and what must be true given those laws. Biology, on the other hand, is the science of the possible, investigating processes that are possible, given those fundamental laws, but not necessary.

Biology is consequently much harder than physics but also infinitely richer in its potential, not just for understanding life and its history but for understanding the universe and its future. The past belongs to physics, but the future belongs to biology.

STUART KAUFFMAN: Chris Langton is wild, scattered, deeply intuitive, not very critical, very creative, good, good, good intuitions. His thesis about the edge of chaos, phase transitions, was a lovely thing to have done. He developed the ideas of looking at cellular automata and a phase transition, and the idea that you have to ask how complex systems can actually generate, create, and pass information around. The idea that it may happen best as a phase transition may not be correct, but it's a lovely hypothesis.

J. DOYNE FARMER: Chris Langton is a cellular-automata engineer. He's following in the footsteps of John von Neumann, engineering self-reproducing systems with cellular automata. Von Neumann showed that it was possible to hand-engineer a self-replicating pattern—engineer a universe—by writing down a particular cellular automaton with a certain set of rules. He was able to show that there are patterns in that universe that replicate themselves, and are both construction-universal and computation-universal. What that means is that the pattern is capable of constructing any pattern, and, secondly, that it's capable of making any computation that a computer can do.

This is a profound step, from an intellectual point of view, but so far it hasn't had any practical consequences. It has the potential for practical consequences—for example, a NASA project investigated the possibility of self-replicating lunar-mining modules based on von Neumann's automaton. But this was never realized, in part because there are still major problems left to be solved, reflecting deficiencies in von Neumann's original work. Von Neumann's automaton has some of the properties of a living system, but it is still not alive. To make an automaton that is really alive, there are a lot of important questions that still need to be answered, both from an engineering and a theoretical point of view. Real organisms do more than just reproduce themselves; they also repair themselves. Real organisms survive in the noisy environment of the real world. Real organisms were not set in place, fully formed, hand-engineered down to the smallest detail, by a conscious God; they arose spontaneously through a process of self-organization. To accomplish von Neumann's original goals, all these problems need to be solved. Von Neumann has given us hope that we can show that life is an abstract, logical process, but until these problems are solved, the demonstration is incomplete. My belief is that the solution lies not in the details but rather in the need for a fundamentally different approach.

The demonstration of a purely logical living system, existing only in an abstract mathematical world, is the goal that Chris and others are working toward. If they succeed, then we will have a new and profound understanding of life.

RICHARD DAWKINS: I met Chris Langton at the first artificial-life conference, at Los Alamos in 1987, which he organized and to which he invited me. Very interesting man, immensely energetic and stimulating, very good at seeing connections between what different people are doing, very good at getting people together who spark off each other in interesting ways. A thoroughly good influence on science.

W. DANIEL HILLIS: Chris Langton is the central guru of this artificial-life stuff. He's onto a good idea when he says that life seems to be at the transition between order and disorder, as he calls it: right at the edge of chaos, just at the temperature between where water is ice and where water is steam, that area where it's liquid—right in between. In many ways, we're poised on the edge between being too structured and too unstructured.

Chris is taking a physical way of thinking of things like phase transitions and dynamical systems and applying it to biological organisms. He also has been a good cheerleader for the whole field, and he's correctly taking the view that it's important not to narrow the field too quickly, even if we get a lot of bad ideas right now. The field of artificial life, for better or for worse, is very inclusive at the moment and includes a lot of junk but also a lot of good stuff. Chris deserves much of the credit for that.

DANIEL C. DENNETT: Chris has a wonderful talent for helping people see what their ideas are and what they aren't; as a midwife of ideas, he's very good. He's playing an important role in artificial life right now. There are so many different ways of doing it, and people are so passionate about what they think the right way is, that it's a good thing there's somebody like Chris around who's not completely wrapped up in any one of those visions to the point where he can't see what people are talking about. He's very good at seeing what people are talking about.

FRANCISCO VARELA: I disagree with his reading of artificial life as being functionalist. By this I refer to his idea that the pattern is the thing. In contrast, there's the kind of biology in which there's an irreducible side to the situatedness of the organism and its history, whether individual or phylogenetic.

Functionalism was a great tradition in artificial intelligence; it's what early AI was all about. In biology, and in AI, there's now a revolution against that, as seen in the work of people like the MIT robotician Rodney Brooks. This revolution leads us to conclude that the way in which a particular style of living develops is inseparable from the fact that it has constructed regularities that operate just to create a viable story. For example, in Stephen Jay Gould's *Wonderful Life,* there are innumerable ways in which life could have developed on Earth, and the fact that some forms of life remain and some others don't is not some kind of optimum model of this or that but fundamentally and intrinsically a historical phenomenon. In this sense, Chris is of the old guard—of the functionalist school.

MURRAY GELL-MANN: Chris Langton is a very interesting researcher. First of all, he was associated early on with the notion that adaptation may involve the attraction of a system toward a region between order and disorder, a sort of transition region, with certain important features. It may be a favorable regime for adaptation, and it's apparently a regime in which scaling laws are likely to apply. It was mathematical work, by Norman Packard and Langton himself, on so-called cellular automata that led to this conjecture, and further research on those automata may or may not support the idea, but there are now many other reasons to pursue the subject.

Meanwhile, Chris has gone on to stir up interest in what he calls artificial life. I myself don't use that way of slicing things. I believe we can learn most by considering natural and artificial complex adaptive systems together—in my view, they form a single subject. Moreover, I don't myself subscribe to the idea that it's valuable to separate out those artificial systems that imitate organisms or biological evolution in certain respects and put them in a separate category from those that imitate other natural complex adaptive systems, such as human societies. However, Chris's category has caught on in a remarkable way, and the term "artificial life" is now widely used. Chris has managed in that way to attract a great deal of attention to the field of plectics and to draw a lot of people into it.

Lately, he's been working on a general computing technique called SWARM for imitating some of the properties of natural complex adaptive systems. At the moment, it looks very promising.

Chapter 22

J. DOYNE FARMER

"The Second Law of Organization"

W. DANIEL HILLIS: *Doyne was in that group of physicists at Los Alamos who were starting to think about complexity, nonlinear phenomena, and adaptive systems. They began to realize that things like "strange attractors" were really ubiquitous in any kind of system—economic systems and biological systems, not just physical systems. That was an incredibly important idea, because it allowed all these people to start talking to each other.*

• • •

J. DOYNE FARMER *is a physicist, an external professor at the Santa Fe Institute, and a cofounder of Prediction Company, an investment firm.*

J. DOYNE FARMER: In the last half of this century, the view has emerged that life and consciousness are natural and inexorable outgrowths of the emergent and self-organizing properties of the physical world. This fundamental change in our view of consciousness and life gives us a new way of looking at ourselves and our beliefs, and of understanding how we fit into the universe.

Not that this is a fait accompli—it's a story in progress, an evolving idea about which there's no universal agreement. Our scientific understanding is still highly fragmented, and we await major breakthroughs as far as anything resembling broad theories is concerned. There's been little serious discussion of how this new view impacts philosophy or sociology. But it's rapidly taking hold, and the change is profound. More than ever, it's becoming impossible to contemplate seriously any philosophical or social question without understanding recent developments in science.

As a kid, I could never shrug off those nagging "Why" questions. It seemed really important to know why we were here, and to understand the meaning of life. It was upsetting to me that these questions, which seemed to lie at the foundation of everything, didn't have any good answers. The easy solutions just didn't fit. My brief preadolescent foray into religion left me with nothing but the realization that people have a desperate need to understand these questions.

When I arrived at college, I immediately took philosophy, picking it out as the subject where the "Why" questions would receive plenty of attention. But as I learned a little philosophy, I became frustrated by the endless debates that seemed to hinge on the meaning of words that could never be defined. Nothing was ever answered. I decided that "Why" questions are simply too deep to be answered with a frontal attack, using the sloppy weapon of human language. Perhaps I wasn't quite as naive as to have expected answers, but at any rate I wasn't satisfied by the study of philosophy.

Physics, on the other hand, seemed to have plenty of answers but not to the "Why" questions. Where I'd imagined that I would learn the foundations, the big principles that made the universe tick, we were instead memorizing formulas about masses on inclined planes. But somehow, I hoped, we'd eventually get to the good stuff. The masses and inclined planes were just an initiation rite, and in the meantime I might learn something tangible, perhaps even useful.

As I progressed through the physics curriculum, I did begin to learn something about fundamental principles—on my own and in discussion with other students—somewhere between the cracks of the problem sets. There was some satisfaction in this. And on learning more astronomy, in the phenomenon of "averted vision," I found a justification for my rationale about the roundabout path to metaphysics via physics: to see a faint star, it's necessary to look away from it; as soon as one looks at it directly, it vanishes.

But as I approached the end of the physics curriculum, there was still something lacking. Averted vision is all well and good, but it *is* necessary to look roughly in the right direction. Physics, in its quest for simple problems, has traditionally focused entirely on the immediate and direct aspects of matter and energy. What makes things move, what makes them get hot or cold. Pushing, pulling, bumping, smashing, and waving. The material aspect of the world, leading to fundamental ideas such as the curvature of spacetime, the quantum nature of reality, the uncertainty principle. All relevant to the big questions. But, the big questions inevitably hinge upon the nature of life and intelligence. While modern physics may say that science necessarily has a subjective element, it says nothing about the nature or origins of consciousness.

It seemed that fundamental physics was stuck. The particle physicists were smashing particles into each other with ever-increasing force, trying to discover how many quarks could dance on the head of a pin. The cosmologists were working with very few facts, debating different flavors for the universe on what seemed to me to be mainly religious grounds. And most of physics was still focused on pushing and pulling, on the material properties of the universe rather than on its informational properties. By informational properties, I mean those that relate to order and disorder. Disorder is fairly well understood, but order isn't. But I'll come back to this later.

I had the good fortune, in graduate school at UC Santa Cruz, to come into contact with some exceptional thinkers: my fellow graduate students Jim Crutchfield, Norman Packard, and Robert Shaw. We spent a lot of time hanging out together, thinking, talking, and sharing our ideas about just about everything. We mused about the informational properties of nature, and the natural origins of organization, and our discussions had a lot of influence on my thinking about these questions.

Norman and I had been friends from childhood, back in Silver City, New Mexico, and we'd always dreamed of starting a company together. So when I had a convenient break in my studies—having passed my qualifying exam and done all my course work, and being a bit dissatisfied with where my research into galaxy formation in unusual cosmologies was going—I decided to take a year off to work with Norman and some other friends, following up on an idea of Norman's. The scheme was to use Newton's laws to beat roulette: in experiments done in our basement, we determined that by means of a computer concealed in the soles of our shoes and activated by a toe switch, we could measure the velocity of the roulette ball and wheel and predict the ball's landing position. Thus ensued a wild time of desperate and adventurous living. The basic idea worked—we made some money in the casinos—but the problems of doing this regularly enough and at sufficiently high stakes prevented us from making very much money. Scientifically, it forced me to learn all about computers (we built what may have been the first concealable digital computers), and it gave me a deep appreciation for the problem of prediction and the curious way in which an apparently simple physical system could be very difficult to predict.

So when Rob Shaw showed up one day and started talking about the phenomenon of "chaos," which he had just learned about, the idea had immediate relevance for me. I instantly understood what he was talking about, and why chaos was important to physical systems like roulette wheels. Rob, Norman, Jim, and I banded together to form the Dynamical Systems Collective at Santa Cruz, and all of us ended up doing our dissertations on the subject of chaos, using one another as our primary thesis advisors. We had a lot of fun doing it.

The fascination with chaos is that it explains some of the disorder in the world, how small changes at one time can give rise to very large effects at a future time. And it shows how simple mathematical rules can give rise to complicated behavior. It explains why simple things can be hard to predict—so much so that they appear to behave randomly. I was lucky enough to get involved in chaos theory fairly early on, and it was great to be in a field that was sufficiently undeveloped that there were a lot of easy problems lying around to be solved.

As I finished graduate school, I really wasn't very sure about getting a job. I'd never been keen on the idea of traditional jobs anyway, and with a degree in "chaos," which at the time very few people had heard of, and no advisor to argue my case, the prospect of a job in science seemed pretty remote. But I happened to see a poster soliciting applications for the Oppenheimer Fellowship at Los Alamos National Laboratory. I'd just been reading about Oppenheimer, and Los Alamos was in New Mexico, where I was raised and where I wanted to return, so even though I was very suspicious of the idea of working at a weapons laboratory I applied for the fellowship. I flew out for a visit, and I was immediately impressed. The people there were exciting, enthusiastic, intelligent, and scientifically they were anything but conservatives. They didn't care at all if what I was doing was not traditional physics. There was a tradition of intellectual freedom there that I haven't seen anywhere else. I ended up with a joint position, split between the Center for Nonlinear Studies and the Theoretical Division. They immediately gave me a lot of responsibility and resources and also gave me carte blanche to do whatever I wanted. Visitors streamed through from all over the world, studying everything under the sun, well beyond the traditional boundaries of physics and mathematics, and I learned an enormous amount just by listening and asking questions.

I continued my work on chaos, but as time went on I began to get a little bored, and increasingly began to think about how to get a handle on the opposite problem: Why is the universe so organized? In 1983 the Center for Nonlinear Studies provided some money for a conference on cellular automata, which I organized with Tomas Toffoli and Stephen Wolfram, and in 1986 Alan Lapedes, Norman Packard, Burton Wendroff, and I organized a conference on "Evolution, Games, and Learning." These conferences were a lot of fun, and

gave us a chance to invite people working on all sorts of crazy, fascinating, and obscure things—simulating life in computer worlds, and
so forth. These conferences put us in contact with the then tenuous
network of people interested in these kinds of things, and that's how I
got to know people such as Chris Langton and John Holland. There
were others at Los Alamos working on related topics; Alan Lapedes
and Dave Sharp were working on neural nets, people in the Theoretical Biology group were working on informational studies of DNA and
also some very interesting aspects of the immune system. We were
able to hire some really good postdocs interested in self-organization,
like Steen Rasmussen and Walter Fontana, and in 1988 we started
the Complex Systems group. Meanwhile, the Santa Fe Institute was
just getting started, which brought in even more interesting people
and expanded the horizon to include subjects, such as economics,
that we hadn't paid much attention to.

Around 1986, Norman Packard and I got involved in two related
projects: one with Alan Perelson involving a simulation of learning
self/nonself recognition and evolution in the immune system, and the
other with Stuart Kauffman, which was a simulation of prebiotic evolution. The idea of the simulation was similar in both cases: we made
up some rules that allowed the parts of the system to evolve and interact with each other. In the case of the immune system, the parts
were concentrations of different kinds of antibodies. For prebiotic
evolution, they were concentrations of molecules such as proteins;
the purpose was to show how a metabolism could arise spontaneously, without the presence of self-replicating molecules like DNA.
The interesting and novel aspect of both simulations was that as the
systems evolved, the compositions of their parts, and hence the
parts' interactions, changed. This all came out of a few simple rules.
We didn't have to put in anything by hand, other than the basic laws
of chemistry—or our crude approximations of them.

The problems turned out to be harder than we'd originally hoped,
and our early results weren't very conclusive. In the case of the autocatalytic networks, I was lucky to have a graduate student, Rik
Bagley, who had migrated from San Diego and badly wanted to produce a Ph.D. thesis. Rik worked hard and ended up getting some nice
results that showed there was real value in the whole approach.

To understand what we did, you first have to understand one of the basic questions relating to the origin of life. Speaking crudely, a living system—an organism—consists of a symbiotic relationship between a metabolism and a replicator. The metabolism, which is built out of proteins and other stuff, extracts energy from the environment, and the replicator contains the blueprint of the organism, with the information needed to grow, make repairs, and reproduce. Each needs the other: the replicator contains the information to make the proteins, the RNA, and other molecules that form the metabolism and run the organism; and the metabolism provides the energy and raw materials needed to build and run the replicator.

The question is, How did this "I'll scratch your back, you scratch mine" situation ever get started? Which came first, the metabolism or the replicator? Or can neither exist without the other, so that they had to evolve together?

In the 1950s, the chemist Harold Urey and the biologist Stanley Miller showed that it was possible for the basic building blocks of proteins—amino acids—to form spontaneously from "earth, fire, and water." However, the synthesis of the more complicated molecules needed in order to form replicators and metabolisms was much less clear. We were trying to demonstrate that a metabolism could spontaneously emerge from basic building blocks and evolve without the presence of a replicator. That is, it could be its own replicator, with the information stored simply in the so-called primordial soup. Starting with simple components—for example, simple amino acids—we wanted to get complex proteins: that is, long, highly diverse chains of amino acids. The basic principle of an autocatalytic network is that even though nothing can make itself, everything in the pot has at least one reaction that makes it, involving only other things in the pot. It's a symbiotic system, in which everything cooperates to make the metabolism work—the whole is greater than the sum of the parts. If normal replication is like monogamous sex, autocatalytic reproduction is like an orgy. We were interested in the logical possibility for this to happen—in an artificial world, simulated inside a computer, following chemical laws that were similar to those of the real world but vastly simplified to make the simulation possible.

In our first simulation, not much happened. The soup of amino acids pretty much remained just that. But after several years of work,

Rik managed to speed up the simulation by a factor of 100 and expand things so that the chemistry was considerably more realistic. As we added features and understood the system better, we began to see things happening. We found that by setting the parameters of the system—which you can think of as determining the relative amounts of earth, fire, and water—we were able to set the knobs so that during the simulation the soup would spontaneously transform itself into a complex and highly specific network of large molecules. Not all molecules. Even though billions of different types are possible, only tens to hundreds are produced. This is like a real metabolism. Furthermore, in some work with Walter Fontana, we were able to show that the system could evolve: new "proteins" would emerge spontaneously, competing with the ones that were already there and changing the metabolism.

What we did in simulating the spontaneous emergence of evolving autocatalytic metabolisms is just one example of an approach that people like Chris Langton, Danny Hillis, and others are taking these days to study the evolution of complex systems. Physics has made most of its breakthroughs when it was able to find simple systems that capture the essence of something, without all the complications. One of the keys to understanding quantum mechanics was the hydrogen atom—the simplest atom—where the mathematics of quantum mechanics could be solved and its consequences understood.

The goal is to find a simple evolving system that contains some of the essential properties of evolving complex systems in general, but without all the complications of the real world. The other goal is to find *lots* of *different* evolving complex systems, and to try to determine what's common to all of them. What is the *essence* of what makes them complex? But at this point we still understand very little. Everyone's still arguing about what a "complex system" really is, and what "organization" means, and whether evolution really tends toward states of greater organization.

For many of us, the goal is to find what might be called "the second law of self-organization." The "second law" part is thrown in as a kind of joke; it's a reference to the second law of thermodynamics, which states that there's an inexorable tendency toward entropy—that is, for physical systems to become disordered. The paradox that immediately bothers everyone who learns about the second law is this: If systems tend to be become more disordered, why, then, do we see so much order around us? Obviously there must be something else going on. In particular, it seems to conflict with our "creation myth": In the beginning, there was a big bang. Suddenly a huge amount of energy was created, and the universe expanded to form particles. At first, things were totally chaotic, but somehow over the course of time complex structures began to form. More complicated molecules, clouds of gas, stars, galaxies, planets, geological formations, oceans, autocatalytic metabolisms, life, intelligence, societies. . . . If we take any particular step in this story, with enough information we can understand it without invoking a general principle. But if we take a step back, we see that there's a general tendency for things to get more organized no matter what the particular details are. Perhaps not everywhere, just in some places at some times. And it's important to stress that no one is saying the second law of thermodynamics is wrong, just that there is a contrapuntal process organizing things at a higher level.

One view of this, perhaps the mainstream view, is that everything depends on a set of disconnected "cosmic accidents." The emergence of organization in the universe depends on a series of highly unlikely unrelated details. The emergence of life is an accident, unrelated to the emergence of all the other forms of order we see in the universe. Life can occur only if all the physical laws are exactly as they are in our universe, and when conditions are almost exactly as they are on our planet.

Many of us find this view implausible. Why would so many different types of order occur? Why would our situation be so special? It seems more plausible to assume that "accidents tend to happen." An individual automobile wreck may seem like a fluke—in fact, *most* automobile wrecks may seem like flukes—but on average they add up. We expect a certain number of them to happen. Our feeling is that the progression of increasing states of organization in the evolution from clouds of gas to life is not an accident. What we want to do is understand the common thread in the pattern, the universal driving force that causes matter to spontaneously organize itself.

This point of view isn't new. It was articulated in the nineteenth century by Herbert Spencer, who wrote about evolution before Darwin and who coined the terms "survival of the fittest" and "evolution." Spencer argued in a very articulate way for the commonality of these processes of self-organization, and used his ideas to make a theory of sociology. However, he was not able to put these ideas into mathematical form or argue them from first principles. And no one else has, either—doing so is perhaps *the* central problem in the study of complex systems.

Many of us believe that self-organization is a general property—certainly of the universe, and even more generally of mathematical systems that might be called "complex adaptive systems." Complex adaptive systems have the property that if you run them—by just letting the mathematical variable of "time" go forward—they'll naturally progress from chaotic, disorganized, undifferentiated, independent states to organized, highly differentiated, and highly interdependent states. Organized structures emerge spontaneously, just by letting the system run. Of course, some systems do this to a greater degree than others, or to higher levels than others, and there will be a certain amount of flukiness to it all. The progression from disorder to organization will proceed in fits and starts, as it does in natural evolution, and it may even reverse itself from time to time, as it does in natural evolution. But in an adaptive complex system, the overall tendency will be toward self-organization. Complex adaptive systems are somewhat special, but not extremely special; the fact that simple forms of self-organization can be seen in many different computer simulations suggests that there are many "weak" complex adaptive systems. A weak system gives rise only to simpler forms of self-organization; a strong one gives rise to more complex forms, like life. The distinction between weak and strong may also depend on scale: even though something like Danny Hillis's "connection machine" is big, it's nothing compared with the Avogadro's number of processors that nature has at her disposal.

Of course, almost none of this is very well understood at this point. That's part of the challenge and fun of thinking about it! We don't know what "organization" is, we don't know why some systems are adaptive and some aren't, we don't know how to tell in advance whether a system is weakly or strongly adaptive, or whether there's a minimum degree of complexity that a system has to have in order to be adaptive. We do know that complex adaptive systems have to be nonlinear and capable of storing information. Also, the parts have to be able to exchange information, but not too much. In the physical world, this is equivalent to saying that they have to be at the right temperature: not too hot, not too cold.

Many simulations show this—in fact, finding the right temperature was one of the breakthroughs in our simulation of autocatalytic metabolisms. We know a little bit about what distinguishes an adaptive complex system from a nonadaptive complex system, such as a turbulent fluid flow, but most of this is lore—anecdotal evidence based on a few observations and cast in largely vague and undefined terms.

To return to the question of who and what we are: if you accept my basic theme that life and intelligence are the result of a natural tendency of the universe to organize itself, then we are just a passing phase, a step in this progression. Of course, one has to be very careful in generalizing from one level of evolution to another. One of the factors that caused Spencer's ideas to lose popularity was social Darwinism—the idea that those who were wealthy and powerful had become that way because they were somehow naturally "fit," while the downtrodden were unfit—which was a poor extension from biological to social evolution, based on a simpleminded understanding of how biological evolution really works. Social evolution *is* different from biological evolution: it's faster, it's Lamarckian, and it makes even heavier use of altruism and cooperation than biological evolution does. None of this was well understood at the time.

Another logical consequence of the evolutionary view is that humans aren't the endpoint of the process. Everything is evolving all the time. At this point, we happen to be the only organism with a sufficiently high degree of intelligence to be able to control our environment in a major way. That gives us the capability to do something remarkable—namely, change evolution itself. If we choose, we can use genetic engineering to alter the character of our offspring. As we understand the details of the human genome better, we're almost certain to do this in order to prevent disease. And we'll be tempted to go beyond that, and increase intelligence, say. There'll be an enormous debate, but with overpopulation, a decreased need for unskilled and manual labor, and pressure from cybernetic intelligences, the motivation to do this will eventually become overwhelming.

Cybernetic intelligences are a consequence of the view that self-organization and life are the natural outcomes of evolution in an adaptive complex system. We're rapidly creating an extraordinary, silicon-based petri dish for the evolution of intelligence. By the year 2025, at the present rate of improvement of computer technology, we're likely to have computers whose raw processing power exceeds that of the human brain. Also, we're likely to have more computers than people. It's difficult to realistically imagine a world of cyberintelligences and superintelligent humanlike beings. It's like a dog trying to imagine general relativity. But I think such a world is the natural consequence of adaptive complex systems. What's even more staggering is that it's not so far in the future—I would say a hundred years at the maximum. One of the amazing features of evolution is that it happens faster and faster. This is particularly vivid in the evolution of societies. Once we can manipulate our own genome, Lamarckian fashion, the rate of change will be staggering.

As for myself, I'm just going along, trying to stay sane, raise my children, and make a living. By July 1991, I'd become fed up with Los Alamos. When I became a group leader, I came fully into contact with the political struggles required to maintain funding. The winding down of the Cold War, combined with increased congressional scrutiny, increased bureaucracy, and poor management, made things tough at Los Alamos. The lab funds basic research by imposing a tax on all the money that comes in and then redistributing it. As the Cold War warmed up, weapons funding went down, the internal tax revenues went down, and basic research became a desperate, survival-oriented enterprise. The Golden Age of science at Los Alamos was over, or at least on hold. The Cold Warriors who used to build weapons were now fiercely scavenging for funds, making up for their lack of skill in science with skill in politics and the urge to survive. Meanwhile, Congress felt that this discretionary tax was subverting their control over the way scientists spend taxpayers' money, and increasingly channeled money into micromanaged Big Science funding initiatives. Running a group in an avant-garde, unestablished area was not fun anymore, depending more on political acumen and fund-raising skills than scientific ability.

So I quit my job at Los Alamos and joined up again with my old friend Norman Packard to take another shot at the global casino. We rounded up some venture capital, recruited another ex-graduate student in physics at UC Santa Cruz, Jim McGill, to run the business side of things, and started Prediction Company, in Santa Fe. Our goal is to make money by predicting and trading in financial markets.

Prediction Company is in part an outgrowth of our work in chaos. One of the reasons for being interested in chaos to begin with is that it presents the possibility that something that seems random may have some underlying simplicity, which can be exploited to make better predictions. In 1987 Sid Sidorowich and I wrote a paper showing how to exploit the order underlying chaos, so that some forms of chaos could be predicted without knowing anything about the underlying dynamics, by building models based only on historical data. We applied this to several phenomena, like fluid flow, sunspots, and ice ages, and got some reasonable results.

It turns out that predicting financial markets doesn't have a lot to do with what Sid and I wrote about earlier, but some of the same techniques work. At Prediction Company, we gather data about financial markets, like currency exchange rates. We apply our learning algorithms to the data, looking for patterns that seem to persist through time. We build models that make trades based on these patterns, and implement them. Every day, data flows into Santa Fe from all over the world, triggers our computer programs to make predictions and trades, which are then sent around the world to the appropriate financial markets. It takes about a minute from the time we receive the data until the trade gets made. So far so good. We have a nice contract with the Swiss Bank Corporation—they provide us with money to trade with, advance us money to pay our bills as needed—and we get a cut of the profits. We're just ramping up to the point where we're trading enough money to make a significant profit. So in the next few years we should either sink or swim.

If we succeed, it will show that, contrary to mainstream theories in economics, it's possible to beat the market. Our feeling is that one of the main causes for the patterns we find is mass psychology: traders respond to information in a predictable manner. So if we can predict the market, and our feeling is right, then it shows that the behavior of groups of human beings is predictable. We're not basing our predictions on a fundamental theory about human nature, but rather on patterns and data. Time will tell whether or not we're right.

These days, scientists are largely treated like beggars, their tin cups eternally extended to government funding agencies. If we succeed, then I'll have the luxury of being able to be a scientist without

having to be a beggar. I hope to get back into the fray of pure research in complex systems before I'm too old and senile to think
clearly anymore. There are some big questions to be answered,
which can give us significant hints about the meaning of life. I'd like
to get back on the front lines of answering these questions.

FRANCISCO VARELA: Doyne Farmer comes from the pure-mathematics
tradition. He's one of the best examples of somebody who took the
very abstract theory of dynamical systems and chaos theory and
brought it down to a concrete level, where you can put it to work in
interesting ways. For example, he's made concrete applications in
economics. He's demonstrated that you can make short-term predictions about phenomena that are intrinsically chaotic, intrinsically
random-looking. That's a major contribution, and in that sense he's
quite an impressive applied mathematician.

Doyne is somebody who has stayed away from the Santa Fe Institute hype as he pursues his work. Since his work is so fundamental
to the Santa Fe project, his name figures. His reputation carries far
beyond the institute because his work and persona are so unique and
interesting that he's been one of the leading characters in several recent books—some of them best-sellers—by science journalists.

Everybody knows what he and Norman Packard are doing at the
Prediction Company, but nobody knows exactly how well or how
badly they're doing it. If you have a few percent more accuracy than
the best intuitive guesses of the good players on Wall Street, you still
stand to make gazillions of dollars—for a while, until everybody else
figures what you're doing. That will give them a window of a year or
two, probably, which is enormous.

BRIAN GOODWIN: The first time I met Doyne Farmer was at Los
Alamos, and he impressed me as somebody who is fantastically on
the ball. They were working on that origin-of-life scenario, with the
autocatalytic-set story. I found Doyne to be very quick, smart, and
tuned-in to these problems. He's one of the high flyers. It's a pity he
dropped out, but never mind; he's doing what he wants to do.

W. DANIEL HILLIS: Too bad that Doyne Farmer went off and started
his company, because he stopped talking about the good stuff he was

doing. He's trying to use it to get rich in the stock market. Doyne is one of the few people I know who's really good explaining physical ideas to people in other fields.

Doyne was in that group of physicists at Los Alamos who were starting to think about complexity, nonlinear phenomena, and adaptive systems. They began to realize that things like "strange attractors" were really ubiquitous in any kind of system—economic systems and biological systems, not just physical systems. That was an incredibly important idea, because it allowed all these people to start talking to each other. It allowed Stuart Kauffman to make bridges between biology and physics. It allowed people like the economist Brian Arthur to make bridges between economics and biology. And in some sense it provided a context, a set of ideas that were discipline-independent, which was very important.

MURRAY GELL-MANN: Doyne Farmer is a very bright scientist, originally a theoretical physicist. He spent a long time at Los Alamos National Laboratory, doing excellent work at the Center for Nonlinear Studies. He was one of the people who really got the CNLS excited about branching out from chaotic phenomena in physics into much more general interests, including the study of complex adaptive systems of many kinds. A number of the people who attended the CNLS meeting on evolution, learning, and games have subsequently become involved with the work of the Santa Fe Institute.

Then he and Norman Packard decided they'd go from research into founding an investment firm, utilizing their discoveries about the not entirely random character of the fluctuations of prices in financial markets. Some dogmatic neoclassical economists had kept claiming that the fluctuations around so-called fundamentals in financial markets amounted to a random walk, and they had produced some evidence to support their assertion. But in the last few years it has been shown—I believe quite convincingly—that in fact various markets show fluctuations that are not entirely random. They're at least partly pseudorandom, and that pseudorandomness can be exploited. The possibility of exploitation depends, of course, on how big a space is being traced out by the nonrandom aspects of these fluctuations—as measured, for instance, by the so-called Hausdorff dimension. If that dimension is too large, then the nonrandomness is very hard to exploit. If the dimension is small, then you can probably make use of it.

They concluded that they could make money using the nonrandomness, and they founded an investment firm based on that idea. For a number of months, they worked with play money, and were quite successful with it, and at that point a financier in Chicago connected them with a Swiss bank, which allowed them to use real money. So far, I believe, it's going pretty well.

RICHARD DAWKINS: I met Doyne Farmer in 1987, at the artificial-life conference organized by his colleague Chris Langton. I've also read *The Eudemonic Pie,* by Thomas A. Bass, and was very amused and entertained by the exploits of Farmer and his friends. Very interesting man.

STUART KAUFFMAN: I've known Doyne since 1984. He's an extremely bright young physicist. Doyne is charismatic, quite brilliant, creative. He did a lot when he was at Santa Cruz to push the early stages of the development of chaos theory.

After Santa Cruz, he came to Los Alamos, where he continued to develop the theory of chaos. By the early 1980s, he realized that chaos was a done deal, that people had done interesting things and it was time to move on to what was becoming the early stage of complexity. He and I, along with Norman Packard, joined forces to work on a model of autocatalytic sets of polymers. Doyne has gone on to think about other things—in particular, the time-series things he's doing now. He's always insightful, always inventive, freewheeling, eclectic, very clever.

Doyne did major things with chaos theory. It's too bad that he's gone off into business. Doyne could pull off a major coup intellectually, so it makes me sad that he's not lending his intuitive inventiveness more to complexity, because he would have a lot to contribute.

CHRISTOPHER G. LANGTON: Doyne Farmer has been a scientific mentor and a good friend, although I don't see him as much as I'd like to these days. Doyne's talents were wasted at Los Alamos, and he had the foresight to escape from LANL and start his own company to apply his nonlinear time-series forecasting techniques to currency and other financial markets. His philosophy is that if his approaches work, they should be self-funding, so he doesn't have to convince some bonehead in Washington that they should be funded. His long-term goal is to make a lot of money in the financial markets, with which he would fund his own institute for the study of complex systems and artificial life. I wish him the best of luck—in a completely objective way, of course.

Part Five

SOMETHING THAT GOES BEYOND OURSELVES

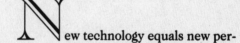

New technology equals new perceptions. As we create tools, we re-create ourselves in their image. Newtonian mechanics gave birth to the metaphor of the heart as a pump. A generation ago, with the advent of cybernetics, information science, and artificial intelligence, we began to think of the brain as a computer. We now have arrived at a new intersection of the empirical and the epistemological. Recent technological breakthroughs in the realm of massively parallel computers and their associated algorithms are having a major impact on the images we have of ourselves and our place in the universe. We have broken through the von Neumann bottleneck of the serial computer.

W. Daniel Hillis brings together, in full circle, many of the ideas in this book: Marvin Minsky's society of mind; Christopher G. Langton's artificial life; Richard Dawkins' gene's-eye view; the plectics practiced at Santa Fe. Hillis developed the algorithms that made possible the massively parallel computer. He began in physics and then went into computer science—where he revolutionized the field—and now he has begun to bring his algorithms to bear on the study of evolution. He sees the autocatalytic effect of fast computers, which lets us design better and faster computers faster, as analogous to the evolution of intelligence. At MIT in the late seventies, Hillis built his "connection machine," a computer that makes use of integrated circuits and, in its parallel operations, closely reflects the workings of the human mind. In 1983, he spun off a computer company called Thinking Machines, which set out to build the world's fastest supercomputer by utilizing parallel architecture.

The massively parallel computational model is critical to the whole set of ideas presented in this book. Hillis's computers, which are fast enough to simulate the process of evolution itself, have shown that programs of random instructions can, by competing, produce new generations of programs—an approach that may well lead to the first machine that truly "thinks." Hillis's work demonstrates that when systems are not engineered but instead allowed to evolve—to build themselves—then the resultant whole is greater than the sum of its parts. Simple entities working together produce some complex thing that transcends them; the implications for biology, engineering, and physics are enormous.

Chapter 23

W. DANIEL HILLIS

"Close to the Singularity"

MARVIN MINSKY: *Danny Hillis is one of the most inventive people I've ever met, and one of the deepest thinkers. He's contributed many important ideas to computer science—especially, but not exclusively, in the domain of parallel computation. He's taken many algorithms that people believed could run only on serial machines and found new ways to make them run in parallel— and therefore much faster. Whenever he gets a new idea, he soon sees ways to test it, to build machines that exploit it, and to discover new mathematical ways to prove things about it. After doing wonderful things in computer science, he got interested in evolution, and I think he's now on the road to becoming one of our major evolutionary theorists.*

• • •

W. DANIEL HILLIS *is a computer scientist; cofounder and chief scientist of Thinking Machines Corporation; holder of thirty-four U.S. patents; editor of several scientific journals, including* Artificial Life, Complexity, Complex Systems, *and* Future Generation Computer Systems; *author of* The Connection Machine *(1985).*

W. Daniel Hillis: I like making things that have complicated behaviors. The ultimate thing that has a complicated behavior is, of course, a mind. The Holy Grail of engineering for the last few thousand years has been to construct a device that will talk to you and learn and reason and create. The first step in doing that requires a very different kind of computer from the simple sequential computers we deal with every day, because these aren't nearly powerful enough. The more they know, the slower they get—as opposed to the human mind, which has the opposite property. Most computers are designed to do things one at a time. For instance, when they look at a picture, they look at every dot in the picture one by one; when they look at a database, they search through the facts one by one. The human mind manages to look all at once at everything it knows and then somehow pull out the relevant piece of information. What I wanted to do was make a computer that was more like that.

It became clear that by using integrated-circuit technology you could build a computer that was structured much more like a human brain; it would do many simple things simultaneously, in parallel, instead of rapidly running through a sequence of things. That principle clearly works in the mind, because the mind manages to work with the hardware of the brain, and the hardware of the brain is actually very slow hardware compared with the hardware of the digital computer.

With modern integrated circuits, it's possible to replicate something over and over again very inexpensively, so I started building a computer by replicating simple processing circuits over and over again and then allowing them to connect with the other interrogatory patterns. Of course, the other thing about your mind is that if I slice up your brain, I see that it's almost all wires. It's all connections between the neurons. Putting into the computer the telephone system that will connect all those little processing elements is the hardest part. That's why my computer was called "the connection machine." I designed it at MIT, but I realized that it was much too big and complicated to be built at a university. It was going to require hundreds of people and tens of millions of dollars. So in 1983 I started the Thinking Machines Corporation, and we spent the next ten years becoming the company that made the world's biggest and fastest computers. The irony is that we were so distracted with all this scientific computing that I haven't made nearly as much progress on the thing I started out with, which is the thinking computer.

My view of what it's going to take to make a thinking machine has changed in recent years. When we started out, I naively believed that each of the pieces of intelligence could be engineered. I still believe that would be possible in principle, but it would take three hundred years to do it. There are so many different aspects to making an intelligent machine that if we used normal engineering methods the complexity would overwhelm us. That presents a great practical difficulty for me; I want to get this project done in my lifetime.

The other thing I've learned is how hard it is to get lots of people to work together on a project and manage the complexity. In some senses, a big connection machine is the most complicated machine humans have ever built. A connection machine has a few hundred billion active parts, all of which are working together, and the way they interact isn't really understood, even by its designers. The only way to design an object of this much complexity is to break it into parts. We decide it's going to have this box and that box and that box, and we send a group of people off to do each of those, and they have to agree on the interfaces before they go off and design their boxes.

Imagine engineering a thinking machine that way. Somebody like Marvin Minsky would say, "O.K., there's a vision box and a reason box and a grammar box," and so on. Then we might break the project up into parts and say, "O.K., Tommy"—Tomaso Poggio, at MIT—"you go off and do the vision box," and we'd get Steve Pinker to do the grammar box, and Roger Schank to do the story box. Then Poggio would take the vision box and say, "All right, we need a depth-perception box, and we need a color-recognition box," and so on.

Then the depth-perception team would say, "O.K., we need a box that perceives depth perception by focus clues and a box that perceives depth perception by binocular vision." Imagine a collection of tens of thousands of people doing these modules, which is how we'd have to engineer it. If you engineer something that way, it has to decompose, and it has to go through these fairly standardized interfaces. There's every reason to believe that the brain is not, in fact, that neatly partitioned. If you look at biological systems in general, while they're hierarchical at a gross level, there's a complex set of interactions between all the parts that doesn't follow the hierarchy. But I'm convinced that our standard methods of engineering wouldn't work very well for designing the brain, although not because of any physical principles we can't control. The brain is an information-processing device, and it does nothing that any universal information-processing device couldn't do.

There's another approach besides this strict engineering approach which can produce something of that complexity, and that's the evolutionary approach. We humans were produced by a process that wasn't engineering. We now have computers fast enough to simulate the process of evolution within the computer. So we may be able to set up situations in which we can cause intelligent programs to evolve within the computer.

I have programs that have evolved within the computer from nothing, and they do fairly complicated things. You begin by putting in sequences of random instructions, and these programs compete and interact with each other and have sex with each other and produce new generations of programs. If you put them in a world where they survive by solving a problem, then with each successive generation they get better and better at solving the problem, and after a few hundred thousand generations they solve the problem very well. That approach may actually be used to produce the thinking machine.

One of the most interesting things is that larger-order things emerge from the interaction of smaller things. Imagine what a multicellular organism looks like to a single-celled organism. The multicellular organism is dealing at a level that would be incomprehensible to a single-celled organism. I think it's possible that the part of our mind that does information-processing is in large part a cultural artifact. A human who's not brought up around other humans isn't a very smart

machine at all. Part of what makes us smart is our culture and our interactions with others. That's part of what would make a thinking
machine smart, too. It would have to interact with humans and be
part of that human culture.

On the biology side, how does this simple process of evolution organize itself into complicated biological organisms? On the engineering side, how do we take simple switching devices like transistors,
whose properties we understand, and cause them to do something
complex that we don't understand? On the physics side, we're studying the general phenomenon of emergence, of how simple things turn
into complex things. All these disciplines are trying to get at essentially the same thing, but from different angles: how can the whole be
more than the sum of the parts? How can simple, dumb things working together produce a complex thing that transcends them? That's
essentially what Marvin Minsky's "society of mind" theory is about;
that's what Chris Langton's "artificial life" is about; that's what
Richard Dawkins' investigation of evolution is about; that's fundamentally what the physicists who are looking at emergent properties
are studying; that's what Murray Gell-Mann's work on quarks is
about; that is the thread that binds all these ideas together.

I am excited by the idea that we may find a way to exploit some
general principles of organization to produce something that goes beyond ourselves. If you step back a zillion years, you can look at the
history of life on Earth as fitting into this pattern. First, fundamental
particles organized themselves into chemistry. Then chemistry organized itself into self-reproducing life. Then life organized itself into
multicellular organisms and multicellular organisms organized themselves into societies bound together by language. Societies are now
organizing themselves into larger units and producing something that
connects them technologically, producing something that goes beyond them. These are all steps in a chain, and the next step is the
building of thinking machines.

To me, the most interesting thing in the world is how a lot of simple, dumb things organize themselves into something much more
complicated that has behavior on a higher level. Everything I'm interested in—whether it's the brain, or parallel computers, or phase
transitions in physics, or evolution—fits into that pattern. Right now,
I'm trying to reproduce within the computer the process of evolution,
with the goal of getting intelligent behavior out of machines. What we

do is put inside the machine an evolutionary process that takes place on a timescale of microseconds. For example, in the most extreme cases, we can actually evolve a program by starting out with random sequences of instructions—say, "Computer, would you please make a hundred million random sequences of instructions. Now, execute all those random sequences of instruction, all those programs, and pick out the ones that came closest to what I wanted." In other words, I defined what I wanted to accomplish, not how to accomplish it.

If I want a program that sorts things into alphabetical order, I'll use this simulated evolution to find the programs that are most efficient at alphabetizing. Of course, random sequences of instructions are unlikely to alphabetize, so none of them does it initially, but one of them may fortuitously put two words in the right order. Then I say to the computer, "Would you please take the 10 percent of those random programs that did the best job, save those, kill the rest, and have the ones that sorted the best reproduce by a process of recombination, analogous to sex. Take two programs and produce children by exchanging their subroutines." The "children" inherit the "traits," the subroutines, of the two programs. Now I have a new generation of programs, produced by combinations of the programs that did a superior job, and I say, "Please repeat that process, score them again, introduce some mutations, and repeat the process again and again, for many generations." Every one of those generations takes just a few milliseconds, so I can do the equivalent of millions of years of evolution within the computer in a few minutes—or, in complicated cases, in a few hours. Finally, I end up with a program that's absolutely perfect at alphabetizing, and it's much more efficient than any program I could ever have written by hand. But if I look at that program, I'm unable to tell you how it works. It's an obscure, weird program, but it does the job, because it comes from a line of hundreds of thousands of programs that did the job. In fact, those programs' lives depended on doing the job.

How do I really know the program will work? In the sorting case, I test it. What if it was something really important? What if this program was going to fly an airplane? Well, you might say, "Gee, it's really scary having a program flying an airplane when we don't have any idea how it works!" But that's exactly what you have with a human pilot; you have a program that was produced by a very similar method, and we have great confidence in it. I have much less confidence in the airplane itself, which was designed very precisely by a lot of very smart engineers. I remember riding in a 747 with Marvin Minsky once, and he pulls out this card from the seat pocket, which said, "This plane has hundreds of thousands of tiny parts, all working together to give you a safe flight." Marvin said, "Doesn't that make you feel confident?"

The engineering process doesn't work very well when it gets complicated. We're beginning to depend on computers that use a process very different from engineering—a process that allows us to produce things of much more complexity than we could with normal engineering. Yet we don't quite understand the possibilities of that process, so in a sense it's getting ahead of us. We're now using those programs to make much faster computers so that we will be able to run this process much faster. The process is feeding on itself. It's becoming faster. It's autocatalytic. We're analogous to the single-celled organisms when they were turning into multicellular organisms. We're the amoebas, and we can't quite figure out what the hell this thing is that we're creating. We're right at that point of transition, and there's something coming along after us.

It's haughty of us to think we're the end product of evolution. All of us are a part of producing whatever is coming next. We're at an exciting time. We're close to the singularity. Go back to that litany of chemistry leading to single-celled organisms, leading to intelligence. The first step took a billion years, the next step took a hundred million, and so on. We're at a stage where things change on the order of decades, and it seems to be speeding up. Technology has the autocatalytic effect of fast computers, which let us design better and faster computers faster. We're heading toward something which is going to happen very soon—in our lifetimes—and which is fundamentally different from anything that's happened in human history before.

People have stopped thinking about the future, because they realize that the future will be so different. The future their grandchildren are going to live in will be so different that the normal methods of planning for it just don't work anymore. When I was a kid, people used to talk about what would happen in the year 2000. Now, at the end of the century, people are still talking about what's going to happen in the year 2000. The future has been shrinking by one year per year, ever since I was born. If I try to extrapolate the trends, to look at where technology's going sometime early in the next century, there comes a point where something incomprehensible will happen. Maybe it's the creation of intelligent machines. Maybe it's telecommunications merging us into a global organism. If you try to talk about it, it sounds mystical, but I'm making a very practical statement here. I think something's happening now—and will continue to happen over the next few decades—which is incomprehensible to us, and I find that both frightening and exciting.

MARVIN MINSKY: Danny Hillis is one of the most inventive people I've ever met, and one of the deepest thinkers. He's contributed many important ideas to computer science—especially, but not exclusively, in the domain of parallel computation. He's taken many algorithms that people believed could run only on serial machines and found new ways to make them run in parallel—and therefore much faster. Whenever he gets a new idea, he soon sees ways to test it, to build machines that exploit it, and to discover new mathematical ways to prove things about it. After doing wonderful things in computer science, he got interested in evolution, and I think he's now on the road to becoming one of our major evolutionary theorists. He's good at telling stories, too. Danny has a terrific mechanical ability. It's not only a sense of knowing how to shape and assemble the necessary materials: he's also one of those rare people who can only be called artists. When he's thinking about how to build something, he can walk through a room of stuff and suddenly notice that several miscellaneous things will fit together perfectly and have the required properties. I'm rather good at that, myself, but Danny is orders of magnitude better.

DANIEL C. DENNETT: I remember first meeting Danny when he was a graduate student at the AI lab at MIT, and it was very clear that he was full of renegade ideas. The one that was particularly strong at the time was the one he's turned into a major contribution—the connection machine. He had the idea of a massively parallel architecture, which would be capable of exploring a different part of the space of possible computations. That opens up a vast area.

What the British mathematician Alan Turing did, with the concept of the Turing machine, was to provide a succinct definition of the entire space of all possible computations. The machine developed by John von Neumann was a mechanical realization of Turing's idea.

A von Neumann machine is the computer on your desk—the standard serial computer. In principle, the von Neumann machine—which is, for all practical purposes, a universal Turing machine—can compute any computable function; but if you don't have a billion years to wait around, you can't actually explore interesting parts of that space. The actual space explorable by any one architecture is quite limited. It sends this vanishingly thin thread out into this huge multidimensional space. To explore other parts of that space, you have got to invent other kinds of architecture. Massive parallel architectures are everybody's first, second, and third choices.

What Danny did was to create if not the first then one of the first really practical, really massive, parallel computers. It precipitated a gold rush. We had a new exploration vehicle, which was looking at portions of design space that had never been looked at before. Danny was very good at selling that idea to people in different scientific fields and demonstrating, with some of the early applications, just how powerful and exciting this vehicle was.

CHRISTOPHER G. LANGTON: Danny Hillis is one of the smartest people I know. I've been influenced by Danny's ideas since his first "AI Memo," from MIT, in which he laid out for the first time his ideas for the connection machine. Danny has a remarkable ability to dive into a new field sight unseen and quickly assess the state-of-the-art of thinking in the field. He can almost immediately work his way to the problems at the cutting edge of the field and make novel and insightful contributions. I sincerely hope he can extricate himself from the business chores of Thinking Machines, Inc., and find the time and support to pursue his scientific work. I have no doubt that he'll come up with significant results.

FRANCISCO VARELA: What I would say about Danny is that he's one of the best exponents of the way of precisely doing complex systems at their best. Not only did he come up with a great invention for a way of doing computation, but he implemented it as a commercial venture. He's done some wonderful work with his connection machine—including, for example, the actual evolution of software by a simulated evolutionary landscape, where programs act like little bugs. They compete, and then there is some kind of selection that produces an optimal code. This is pretty impressive. I'm not sure that it has a lot to do with biological evolution, but it creates an artificial evolution which is very interesting. This is a hybrid way of thinking that's quite extraordinarily imaginative.

There's a trend that I know Danny's thinking about: the notion of inventing artificial worlds and therefore creating parallel universes. We're just at the beginnings of that. So far, in fact, that I think "artificial life" should be called "artificial worlds." Because the interesting part of the idea of taking biological ideas into simulation is to consider the biological entities and their worlds as a complete system— one where you deal with the nonseparation between outside and inside, where you let the biological system actually play out the full game of life, in a world that has as much reality as itself.

ROGER SCHANK: He had a neat idea, and he made a breakthrough in getting massively parallel machines to work. His machines have no impact at all on what I do. There's not a lot of understanding of their value. It's not at all clear why you'd want one. They are a kind of entrepreneurial enterprise, and aren't necessarily practical utilities. I've never seen computing power as the essence in computers; the problem is figuring out what you want to do with them. All that parallel machines do for you is make things a lot faster, but the issue isn't speed; it never has been.

MURRAY GELL-MANN: I'm very fond of Danny Hillis, and I think very highly of him. My impression is that he's not only a daring person, which we know, but also a deep thinker and a very effective one. I wish I knew and understood more about his work. I look forward to seeing more of him, and to learning more about the things that interest him.

SELECTED READING

PART ONE

Cronin, Helena. *The Ant and the Peacock*. New York: Cambridge University Press (1992).

Darwin, Charles R. *On the Origin of Species*. Cambridge: Harvard University Press (1859/1964).

Dawkins, Richard. *The Blind Watchmaker*. New York: W.W. Norton (1986).

_____. *The Extended Phenotype*. New York: Oxford University Press (1982).

_____. *River out of Eden*. New York: Basic Books (1995).

_____. *The Selfish Gene*. 2d ed., New York: Oxford University Press (1989).

Dobzhansky, Theodosius. *Genetics and the Origin of Species*. New York: Columbia University Press (1951).

_____. *Mankind Evolving: The Evolution of the Human Species*. New Haven: Yale University Press (1962).

Eigen, Manfred. *Steps Towards Life*. New York: Oxford University Press (1992).

Eldredge, Niles. *Fossils*. New York: Harry N. Abrams (1991).

_____. *The Miner's Canary*. New York: Prentice Hall (1991).

_____. *The Monkey Business*. New York: Washington Square Press (1982).

_____. *Reinventing Darwin*. New York: John Wiley (1995).

_____. *Time Frames: The Rethinking of Darwinian Evolution and the Theory of Punctuated Equilibria*. New York: Simon & Schuster (1985).

_____. *Unfinished Synthesis*. New York: Oxford University Press (1985).

_____, and Stephen J. Gould. "Punctuated Equilibria: An Alternative to Phyletic Gradualism," in T.J.M. Schopf, ed., *Models in Paleobiology*. San Francisco: Freeman Cooper (1972).

_____, and Marjorie Grene. *Interactions*. New York: Columbia University Press (1992).

Ewald, Paul W. "Cultural Vectors, Virulence, and the Emergence of Evolutionary Epidemiology." *Oxford Surveys in Evolutionary Biology* 5 (1988) 215–245.

Fisher, Ronald A. *The Genetical Theory of Natural Selection*. Oxford: Clarendon Press (1930).

Goldschmidt, Richard. *The Material Basis of Evolution*. New Haven: Yale University Press (1940).

Gould, Stephen Jay. *Bully for Brontosaurus*. New York: W.W. Norton (1992).

_____. *Ever Since Darwin*. New York: W.W. Norton (1977).

_____. *The Flamingo's Smile*. New York: W.W. Norton (1985).

_____. *Hen's Teeth and Horse's Toes*. New York: W.W. Norton (1983).

_____. *The Mismeasure of Man*. New York: W.W. Norton (1981).

_____. *Ontogeny and Phylogeny*. Cambridge: Harvard University Press (1977).

_____. *The Panda's Thumb*. New York: W.W. Norton (1980).

_____. *Time's Arrow, Time's Cycle: Myth and Metaphor in the Discovery of Geological Time*. Cambridge: Harvard University Press (1987).

_____. *An Urchin in the Storm*. New York: W.W. Norton (1988).

_____. *Wonderful Life*. New York: W.W. Norton (1989).

_____, and Richard C. Lewontin. "The Spandrels of San Marco and the Panglossian Paradigm: A Critique of the Adaptationist Programme." *Proc. Roy. Soc. London* B 205 (1979) 581–598.

_____, and Elisabeth S. Vrba. "Exaptation—A Missing Term in the Science of Form." *Paleobiology* 8 (1982) 4–15.

Haldane, J.B.S. *The Causes of Evolution*. London: Longmans (1932).

Hamilton, W.D. "The Genetical Evolution of Social Behaviour" (I and II). *Journal of Theoretical Biology* 7 (1964) 1–52.

_____. "The Moulding of Senescence by Natural Selection." *Journal of Theoretical Biology* 12 (1966) 12–45.

Hull, David L. "Interactors versus Vehicles," in H.C. Plotkin, ed., *The Role of Behavior in Evolution*. Cambridge: MIT Press (1988).

Jones, Steve. *The Language of the Genes: Biology, History, and the Evolutionary Future*. London: HarperCollins (1993).

_____, Robert Martin, and David Pilbeam, eds. *The Cambridge Encyclopedia of Human Evolution*. Cambridge: Cambridge University Press (1992).

Lewontin, Richard C., Steven Rose, and Leon J. Kamin. *Not In Our Genes*. New York: Pantheon (1984).

Lovelock, James. *Gaia*. New York: Oxford University Press (1979).

Margulis, Lynn. *Early Life*. Boston: Jones & Bartlett (1981).

_____. *The Origin of Eukaryotic Cells*. New Haven: Yale University Press (1970).

_____. *Symbiosis in Cell Evolution*. 2d ed., New York: W.H. Freeman (1993).

_____, and Dorion Sagan. *Microcosmos*. New York: Simon & Schuster (1986).

_____. *Mystery Dance*. New York: Summit Books (1991).

_____. *Origins of Sex*. New Haven: Yale University Press (1986).

_____, and Karlene V. Schwartz. *Five Kingdoms: An Illustrated Guide to the Phyla of Life on Earth*. 2d ed., San Francisco: W.H. Freeman (1988).

Maynard Smith, John. *Did Darwin Get It Right?* London: Chapman & Hall (1989).

_____. *Evolution and the Theory of Games*. Cambridge: Cambridge University Press (1982).

_____. *The Evolution of Sex*. Cambridge: Cambridge University Press (1978).

_____. *The Problems of Biology*. Oxford: Oxford University Press (1986).

Mayr, Ernst. *The Growth of Biological Thought*. Cambridge: Harvard University Press (1982).

_____. *Toward a New Philosophy of Biology*. Cambridge: Harvard University Press (1988).

Medawar, Peter. *The Limits of Science*. New York: Harper & Row (1984).

_____. *Memoir of a Thinking Radish*. Oxford: Oxford University Press (1988).

Nesse, Randolph, M.D., and George C. Williams. *Why We Get Sick*. New York: Times Books (1995).

Simpson, George Gaylord. *The Major Features of Evolution*. New York: Columbia University Press (1953).

_____. *The Meaning of Evolution*. New Haven: Yale University Press (1949).

_____. *Tempo and Mode in Evolution*. New York: Columbia University Press (1944).

Snow, C.P. *The Two Cultures*. Cambridge: Cambridge University Press (1993).

Sober, Elliott. *The Nature of Selection*. Cambridge: MIT Press (1984).

Stanley, Steven M. *Children of the Ice Age*. New York: Crown (1995).

_____. *The New Evolutionary Timetable*. New York: Basic Books (1981).

Stebbins, G.L. *Variation and Evolution in Plants*. New York: Columbia University Press (1950).

Sturtevant, A.H. "On the Effects of Selection on Social Insects." *Quarterly Review of Biology* 13 (1938) 74–76.

Thompson, D'Arcy W. *On Growth and Form*. Cambridge: Cambridge University Press (1917).

Trivers, Robert. *Social Evolution*. Menlo Park CA: Benjamin/ Cummings (1985).

Vrba, Elisabeth S., and Niles Eldredge. "Individuals, Hierarchies, and Processes: Towards a More Complete Evolutionary Theory." *Paleobiology* 10 (1984) 146–171.

Waddington, C.H. *The Evolution of an Evolutionist*. Edinburgh: Edinburgh University Press (1975).

_____. *The Nature of Life*. London: Allen & Unwin (1962).

_____. *The Strategy of the Genes*. London: Allen & Unwin (1957).

Weismann, August. "The All-Sufficiency of Natural Selection: A Reply to Herbert Spencer." *Contemporary Review* 64 (1893) 309–338, 596–610.

Williams, George C. *Adaptation and Natural Selection: A Critique of Some Current Evolutionary Thought*. Princeton: Princeton University Press (1966).

_____. *Natural Selection: Domains, Levels, and Challenges*. New York: Oxford University Press (1992).

_____. *Sex and Evolution*. Princeton: Princeton University Press (1975).

Wilson, Edward O. *The Diversity of Life.* Cambridge: Harvard University Press (1992).

_____. *On Human Nature.* Cambridge: Harvard University Press (1978).

_____. *Sociobiology: The New Synthesis.* Cambridge: Harvard University Press (1975).

Wright, Sewall. "Adaptation and Selection," in *Genetics, Paleontology, and Evolution,* G.L. Jepson, E. Mayr, and G.G. Simpson, eds. Princeton: Princeton University Press (1949).

_____. "Evolution in Mendelian Populations." *Genetics* 16 (1931) 97–159.

_____. "Tempo and Mode in Evolution: A Critical Review." *Ecology* 26 (1945) 415–419.

Wynne-Edwards, V.C. *Animal Dispersion in Relation to Social Behaviour.* Edinburgh: Oliver & Boyd (1962).

PART TWO

Chomsky, Noam. *Knowledge of Language.* New York: Praeger (1986).

Dennett, Daniel C. *Brainstorms.* Montgomery VT: Bradford Books (1978).

_____. *Consciousness Explained.* Boston: Little, Brown (1991).

_____. *Content and Consciousness.* London: Routledge & Kegan Paul (1969).

_____. *Elbow Room: The Varieties of Free Will Worth Wanting.* Cambridge: MIT Press/Bradford (1984).

_____. *The Intentional Stance.* Cambridge: MIT Press/Bradford (1987).

Dreyfus, Hubert. *What Computers Can't Do.* 2d ed., New York: Harper & Row (1979).

_____, and S.E. Dreyfus. *Minds Over Matter.* New York: The Free Press (1986).

Hofstadter, Douglas R., and Daniel C. Dennett. *The Mind's I.* New York: Bantam (1982).

Humphrey, Nicholas. *Consciousness Regained.* Oxford: Oxford University Press (1983).

_____. *A History of the Mind.* New York: Simon & Schuster (1992).

_____. *The Inner Eye.* London: Faber & Faber (1986).

_____. "'Interest' and 'Pleasure': Two Determinants of a Monkey's Visual Preferences." *Perception* 1 (1972) 395–416.

Jacob, François. *The Possible and the Actual.* Seattle: University of Washington Press (1982).

Levelt, Willem. *Speaking.* Cambridge: MIT Press/Bradford (1989).

Maturana, Humberto D., and Francisco J. Varela. *Autopoiesis and Cognition: The Realization of the Living.* Boston: D. Reidel (1980).

_____. *The Tree of Knowledge.* Boston: New Science Library (1987).

Minsky, Marvin. *The Society of Mind.* New York: Simon & Schuster (1986).

_____, and Seymour Papert. *Perceptrons.* Rev. ed., Cambridge: MIT Press (1987).

Moravec, Hans. *Mind Children: The Future of Robot and Human Intelligence.* Cambridge: Harvard University Press (1988).

Papert, Seymour. *Mindstorms.* New York: Harper & Row (1981).

Penrose, Roger. *The Emperor's New Mind: Concerning Computers, Minds, and the Laws of Physics.* New York: Oxford University Press (1989).

_____. *Shadows of the Mind: A Search for the Missing Science of Consciousness.* New York: Oxford University Press (1994).

Pinker, Steven. *The Language Instinct.* New York: William Morrow (1994).

_____. *Learnability and Cognition: The Acquisition of Argument Structure.* Cambridge: MIT Press (1989).

_____, and Paul Bloom. "Natural Language and Natural Selection." *Behavioral and Brain Sciences* 13 (1990) 707–784.

Schank, Roger. *The Connoisseur's Guide to the Mind.* New York: Summit Books (1991).

_____. *Tell Me a Story.* New York: Scribners (1990).

_____, and R. Abelson. *Scripts, Plans, Goals, and Understanding: An Inquiry into Human Knowledge Structures.* Hillsdale NJ: Erlbaum (1977).

_____, and Peter Childers. *The Cognitive Computer.* Reading MA: Addison-Wesley (1988).

_____. *The Creative Attitude: Learning to Ask and Answer the Right Questions.* New York: Macmillan (1988).

Searle, John. *Minds, Brains, and Science.* Cambridge: Harvard University Press (1984).

_____. *The Rediscovery of the Mind.* Cambridge: MIT Press (1992).

Varela, Francisco J. *Principles of Biological Autonomy.* New York: Elsevier North Holland (1979).

_____, Evan Thompson, and Eleanor Rosch. *The Embodied Mind.* Cambridge: MIT Press (1992).

PART THREE

Barrow, John D., and Frank J. Tipler. *The Anthropic Cosmological Principle.* New York: Oxford University Press (1986).

Davies, Paul. *The Cosmic Blueprint.* New York: Simon & Schuster (1989).

_____. *The Edge of Infinity.* New York: Simon & Schuster (1981).

_____. *God and the New Physics.* New York: Simon & Schuster (1983).

_____. *The Last Three Minutes: Conjectures about the Ultimate Fate of the Universe.* New York: Basic Books (1994).

_____, ed. *The New Physics.* Cambridge: Cambridge University Press (1989).

_____. "A New Science of Complexity." *New Scientist* 26 November 1988.

_____. *Other Worlds.* London: Dent (1980).

_____. *The Physics of Time Asymmetry.* Berkeley: University of California Press (1974).

_____. *Space and Time in the Modern Universe.* Cambridge University Press, Cambridge, 1977.

_____. *Superforce.* New York: Simon & Schuster (1984).

_____, and John Gribbin. *The Matter Myth.* New York: Simon & Schuster (1992).

Gribbin, John, and Martin Rees. *Cosmic Coincidences: Dark Matter, Mankind, and Anthropic Cosmology.* New York: Bantam (1989).

Guth, Alan. *The Inflationary Universe.* In press.

Hawking, Stephen W. *A Brief History of Time: From the Big Bang to Black Holes.* New York: Bantam (1988).

Pagels, Heinz R. *The Cosmic Code: Quantum Physics As the Language of Nature.* New York: Simon & Schuster (1982).

_____. *Perfect Symmetry.* New York: Simon & Schuster (1985).

Rees, Martin. *Our Home Universe*. In press.

Smolin, Lee. *The Life of The Cosmos: A New View of Cosmology, Particle Physics, and the Meaning of Quantum Physics*. New York: Crown (1995).

Smoot, George, and Keay Davidson. *Wrinkles in Time*. New York: William Morrow (1994).

Weinberg, Steven. *Dreams of a Final Theory*. New York: Pantheon (1992).

_____. *The First Three Minutes: A Modern View of the Origin of the Universe*. Updated ed., New York: Basic Books (1988).

PART FOUR

Farmer, J. Doyne, Tomaso Toffoli, and Stephen Wolfram, eds. "Cellular automata." *Physica* 10D Amsterdam (1984).

_____, eds. "Evolution, Games, and Learning: Models for Adaptation in Machines and Nature." *Physica* 22D Amsterdam (1986).

Gell-Mann, Murray. *The Quark and the Jaguar*. New York: W.H. Freeman (1994).

Holland, John H. *Adaptation in Natural and Artificial Systems*. Ann Arbor: University of Michigan Press (1975).

Kauffman, Stuart A. *Origins of Order: Self-Organization and Selection in Evolution*. New York: Oxford University Press (1993).

_____, and George Johnson. *At Home in the Universe*. New York: Oxford University Press (1995).

Langton, Christopher G., ed. *Artificial Life*. Reading MA: Addison-Wesley (1989).

_____, Charles Taylor, J. Doyne Farmer, and Steen Rasmussen, eds. *Artificial Life II*. Reading MA: Addison-Wesley (1992).

Pagels, Heinz R. *Dreams of Reason: The Computer and the Rise of the Sciences of Complexity*. New York: Simon & Schuster (1988).

Toffoli, Tomaso, and Norman Margolus. *Cellular Automata Machines*. Cambridge: MIT Press (1987).

PART FIVE

Hillis, W. Daniel. *The Connection Machine.* Cambridge: MIT Press (1985).

_____. "Intelligence as an Emergent Behavior," in *Artificial Intelligence,* Stephen Graubard, ed. Cambridge: MIT Press (1988).

INDEX

ABOUT THE AUTHOR

John Brockman, writer and literary agent, is the founder of The Reality Club, president of Edge Foundation, and editor of *EDGE* newsletter. He divides his time between New York City and Bethlehem, Connecticut.